Non-associative Structures and Other Related Structures

Non-associative Structures and Other Related Structures

Special Issue Editor

Florin Felix Nichita

MDPI • Basel • Beijing • Wuhan • Barcelona • Belgrade • Manchester • Tokyo • Cluj • Tianjin

Special Issue Editor
Florin Felix Nichita
Simion Stoilow Institute of
Mathematics of the Romanian
Academy
Romania

Editorial Office
MDPI
St. Alban-Anlage 66
4052 Basel, Switzerland

This is a reprint of articles from the Special Issue published online in the open access journal *Axioms* (ISSN 2075-1680) (available at: https://www.mdpi.com/journal/axioms/special_issues/non-associative_structures).

For citation purposes, cite each article independently as indicated on the article page online and as indicated below:

LastName, A.A.; LastName, B.B.; LastName, C.C. Article Title. *Journal Name* **Year**, *Article Number*, Page Range.

ISBN 978-3-03936-254-7 (Hbk)
ISBN 978-3-03936-255-4 (PDF)

Contents

About the Special Issue Editor

Florin Felix Nichita (Ph.D.—SUNY Buffalo, 2000/2001) is currently Senior Scientific Researcher at the "Simion Stoilow" Institute of Mathematics of the Romanian Academy, and a collaborator of MDPI journals. His primary research interests are in the areas of topology, noncommutative algebra, geometry, etc. With about 60 published works, and many invited talks/visitor positions, his attention has recently been attracted to poetry and literature.

Preface to "Non-associative Structures and Other Related Structures"

Leonhard Euler (1707–1783) was born on 15 April 1707, in Basel, Switzerland. Euler's formula is a mathematical formula in complex analysis that establishes the fundamental relationship between the trigonometric functions and the complex exponential function. When its variable is the number pi, Euler's formula evaluates to Euler's identity.

Bertrand Russell wrote that mathematics can exalt "as surely as poetry". This is especially true of one equation: Euler's identity, the brainchild of Leonhard Euler, the Mozart of mathematics. More than two centuries after Euler's death, it is still regarded as a conceptual diamond of unsurpassed beauty. Called Euler's identity, or God's equation, it includes just five numbers but represents an astonishing revelation of hidden connections. It ties together everything from basic arithmetic to compound interest, the circumference of a circle, trigonometry, calculus, and even infinity. In David Stipp's book, *A Most Elegant Equation: Euler's Formula and the Beauty of Mathematics* (2017), Euler's identity formula becomes a contemplative stroll through the glories of mathematics. The result is an ode to this magical field.

On the other hand, according to Melissa Hogenboom, the most beautiful equation is the Yang-Baxter equation (20 January 2016).

http://www.bbc.com/earth/story/20160120-the-most-beautiful-equation-is-the-yang-baxter-equation.

In this book, we consider connections between Euler's formula and the Yang-Baxter equation.

This discussion includes also dual numbers. Dual numbers and their higher-order variants are important tools for numerical computations, and in particular for finite difference calculus. Based on the relevant algebraic rules and matrix realizations of dual numbers, a novel point of view is presented in this book.

Other interesting topics include non-associative algebras with metagroup relations; branching functions for admissible representations of affine Lie Algebras; super-Virasoro algebras; UJLA structures; etc.

Two more special issues will continue our investigations:

Axioms: Non-associative Structures, Yang–Baxter Equations and Related Topics

Sci: Mathematics and Poetry, with a View Towards Machine Learning

Florin Felix Nichita
Special Issue Editor

axioms

MDPI

Editorial

Non-Associative Structures and Other Related Structures

Florin F. Nichita

Simion Stoilow Institute of Mathematics of the Romanian Academy, 21 Calea Grivitei Street, 010702 Bucharest, Romania; Florin.Nichita@imar.ro

Received: 7 April 2020; Accepted: 10 April 2020; Published: 13 April 2020

In January 2019, MDPI published a book titled *Hopf Algebras, Quantum Groups and Yang–Baxter Equations*, based on a successful special issue. We hope that a book titled *Non-Associative Structures and Other Related Structures* will be published soon.

Non-associative algebras are currently a fashionable research direction. There are two important classes of non-associative structures: Lie structures and Jordan structures. Various Jordan structures play an important role in quantum group theory and in fundamental physical theories. In recent years, several attempts to unify (non-)associative structures have led to interesting results. The UJLA structures are not the only structures that realize such a unification. Associative algebras and Lie algebras can be unified at the level of Yang–Baxter structures.

Several papers published in the open access journal Axioms deal with the Yang–Baxter equation. This equation first appeared in theoretical physics, in a paper (1968) by the Nobel laureate C.N. Yang, and in statistical mechanics, in R.J. Baxter's work (1971). Later, it turned out that this equation plays a crucial role in: quantum groups, knot theory, braided categories, analysis of integrable systems, quantum mechanics, non-commutative descent theory, quantum computing, non-commutative geometry, etc. At the Kyoto International Mathematics Congress (1990), three of the four Fields Medalists were awarded prizes for their work related to the Yang–Baxter equation. Many scientists have used the axioms of various algebraic structures (quasi-triangular Hopf algebras, Yetter-Drinfeld categories, Lie (super)algebras, algebra structures, Boolean algebras, etc.) or computer calculations in order to produce solutions for the Yang–Baxter equation. However, the full classification of its solutions remains an open problem. The Yang–Baxter equation can also be interpreted in terms of logical circuits and, in logic, it represents a kind of compatibility condition when working with many logical sentences in the same time.

It is interesting to note that several special issues published by AXIOMS led to new solutions for the Yang–Baxter equations. In addition, the topics of these special issues were advertised at many conferences (Boston, Bucharest, Brasov, Caen, Galati, Iasi, Malta, Sofia, etc.). Moreover, at the 14th International Workshop on Differential Geometry and Its Applications, hosted by the Petroleum Gas University from Ploiesti, between 9–11 July 2019, the AXIOMS sponsored the "Best Poster Award" for best presented papers to support young scholars (including post-docs up to 35 years old). The winners were announced by the chairs during the workshop.

The authors who contributed to the special issue *Non-associative Structures and Other Related Structures* put in a lot of work in order to write high quality papers: [1–7]. Some other works related to this special issue are the following: [8–12]. Further comments and references for related articles will appear in our next special issues.

Funding: This research received no external funding.

Acknowledgments: We would like thank the authors who contributed to this special issue, the referees and the editorial staff of Axioms.

Conflicts of Interest: The author declares no conflict of interest.

References

1. Behr, N.; Dattoli, G.; Lattanzi, A.; Licciardi, S. Dual Numbers and Operational Umbral Methods. *Axioms* **2019**, *8*, 77. [CrossRef]
2. Kwon, N. Branching Functions for Admissible Representations of Affine Lie Algebras and Super-Virasoro Algebras. *Axioms* **2019**, *8*, 82. [CrossRef]
3. Ludkowski, S.V. Cohomology Theory of Nonassociative Algebras with Metagroup Relations. *Axioms* **2019**, *8*, 78. [CrossRef]
4. Ludkowski, S.V. Smashed and Twisted Wreath Products of Metagroups. *Axioms* **2019**, *8*, 127. [CrossRef]
5. Ludkowski, S.V. Separability of Nonassociative Algebras with Metagroup Relations. *Axioms* **2019**, *8*, 139. [CrossRef]
6. Nichita, F.F. Unification Theories: Examples and Applications. *Axioms* **2018**, *7*, 85. [CrossRef]
7. Nichita, F.F. Unification Theories: New Results and Examples. *Axioms* **2019**, *8*, 60. [CrossRef]
8. Lebed, V. Braided Systems: A Unified Treatment of Algebraic Structures with Several Operations. *Homol. Homotopy Appl.* **2017**, *19*, 141–174. [CrossRef]
9. Oner, T.; Katican, T. On Solutions to the Set-Theoretical Yang–Baxter Equation in Wajsberg-Algebras. *Axioms* **2018**, *7*, 6 [CrossRef]
10. Marcus, S.; Nichita, F.F. On Transcendental Numbers: New Results and a Little History. *Axioms* **2018**, *7*, 15. [CrossRef]
11. Nichita, F. Introduction to the Yang–Baxter Equation with Open Problems. *Axioms* **2012**, *1*, 33–37. [CrossRef]
12. Nichita, F.F. (Ed.) *Hopf Algebras, Quantum Groups and Yang–Baxter Equations*; MDPI: Basel, Switzerland, 2019; ISBN 978-3-03897-324-9 (Pbk); ISBN 978-3-03897-325-6 (PDF).

axioms

MDPI

Article

Dual Numbers and Operational Umbral Methods

Nicolas Behr [1,*], Giuseppe Dattoli [2], Ambra Lattanzi [2,3] and Silvia Licciardi [2]

1 Institut de Recherche en Informatique Fondamentale (IRIF), Université de Paris, Bâtiment Sophie Germain, Case Courier 7014, 8 Place Aurélie Nemours, CEDEX 13, 75205 Paris, France
2 ENEA—Frascati Research Center, Via Enrico Fermi 45, 00044 Rome, Italy
3 H. Niewodniczański Institute of Nuclear Physics, Polish Academy of Science, ul. Radzikowskiego 152, 31-342 Kraków, Poland
* Correspondence: nicolas.behr@irif.fr

Received: 22 May 2019; Accepted: 26 June 2019; Published: 2 July 2019

Abstract: Dual numbers and their higher-order version are important tools for numerical computations, and in particular for finite difference calculus. Based on the relevant algebraic rules and matrix realizations of dual numbers, we present a novel point of view, embedding dual numbers within a formalism reminiscent of operational umbral calculus.

Keywords: dual numbers; operational methods; umbral image techniques

---

1. Introduction

Dual numbers (DNs), introduced during the second half of the 19th century [1–5], can be viewed as abstract entities, similar to ordinary complex numbers, and are defined as

$$z = x + \epsilon y, \qquad (x, y) \in \mathbb{R} \tag{1}$$

where the corresponding "imaginary" unit or *dual number unit (DNU)* ϵ is a nilpotent number,

$$\epsilon^2 = 0 \qquad \text{and} \qquad \epsilon \neq 0. \tag{2}$$

Dual numbers were originally introduced within the context of geometrical studies, and later exploited to deal with problems in pure and applied mechanics [6,7]. For instance, it has been demonstrated in [8–10] how to formulate the equations of rigid body motion in terms of just three "dual" equations instead of their six "real" counterparts (thereby realizing an equivalence between spherical and spatial kinematics). This approach has been extended in [11–13] to a treatment of rigid body motion in terms of a certain variant of "hyper-dual" numbers, implemented in contrast to our approach via sets of "ordinary" dual numbers together with certain algebraic relations that are motivated from the specific requirements within the relevant field of robotics and of mechanics. More recently, as further discussed in the present paper, the importance of a different kind of higher-order dual numbers has been recognized in numerical analysis to reduce round-off errors [14]. We believe that the use of dual numbers in the applied sciences is not as widespread as it could be, and that many new fields of research would benefit from their relevant introduction. An important domain in which they may bring significant novelties is that of the perturbative techniques in classical and quantum mechanics.

The main contribution of this paper consists in fixing the underlying algebraic rules of the dual numbers in the wider context of umbral and operational calculus. The paper is organized as follows: Section 2 delivers a basic mathematical introduction to dual numbers. Section 3 is devoted to the description of the computational procedure based upon dual numbers and umbral calculus. In Section 4, we provide insight into how this powerful method can be applied to deal with problems arising in different contexts. For illustration, we consider the Schrödinger and the heat equation, cornerstones in their respective fields of physics. Section 5 provides a conclusion with further considerations for future works.

2. Higher-Order Dual Numbers

The DN algebraic rules [15,16], summarized below, are a straightforward consequence of the previous identity in Equation (2) (with $z = x + \epsilon y$ and $w = u + \epsilon v$):

$$\begin{aligned}
&\textbf{Component-wise algebraic addition}\\
&z + w = x + u + \epsilon(y + v)\\
&\textbf{Product}\\
&z \cdot w = xu + \epsilon(xv + yu)\\
&\textbf{Inverse}\\
&z^{-1} = \frac{1}{x}\left(1 - \epsilon\frac{y}{x}\right) \quad (x \neq 0)\\
&\textbf{Power}\\
&z^n = x^n\left(1 + n\epsilon\frac{y}{x}\right) \quad (n \in \mathbb{Z}_{\geq 0},\ x \neq 0)
\end{aligned} \tag{3}$$

While the addition operation is entirely analogous to the component-wise addition operation on two-dimensional vectors, the last three operations (product, inverse and power) characterize the distinguishing special algebraic properties of dual numbers (DNs). The multiplication is commutative, associative and distributive, thus DNs form a two-dimensional associative and commutative algebra over the real numbers.

We now extend this traditional dual number formalism as motivated by the following type of problem. Consider the Taylor expansion up to some order k (denoted $\approx_k$) of an at least k-fold continuously differentiable function f around a point x,

$$f(x + y) \approx_k \sum_{m=0}^{k} \frac{y^m}{m!} f^{(m)}(x)\,. \tag{4}$$

Following the *automatic differentiation* paradigm [17–19], since in practice the function f will be implemented in some algorithmic from, it may be advantageous to formulate truncations such as Equation (4) in terms of *generalized (or higher-order) dual numbers*. To this end, let us introduce the families of square matrices ${}_k\hat{\epsilon}_{\pm}$, ${}_k\hat{1}$ and ${}_k\hat{0}$ with entries (for $i, j = 1, \ldots, k$)

$$\left({}_k\hat{\epsilon}_{\pm}\right)_{i,j} := \delta_{j,i\pm 1}\,, \qquad \left({}_k\hat{1}\right)_{i,j} := \delta_{i,j}\,, \qquad \left({}_k\hat{0}\right)_{i,j} := 0\,, \tag{5}$$

where $\delta_{i,j}$ denotes the Kronecker symbol. It is straightforward to verify that for all $k \geq 2$ and $\ell \geq 0$

$$\left({}_k\hat{\epsilon}_{\pm}^{\ell}\right)_{i,j} = \delta_{j,i\pm\ell} \quad \Rightarrow \quad \left({}_k\hat{\epsilon}_{\pm}\right)^k = 0\,. \tag{6}$$

Then, under the assumptions in Equation (4), endowing the function $f(x)$ suitably with a component-wise action on square matrices, we find (for $k \geq 2$)

$$f\left(x\,{}_k\hat{1} + y\,{}_k\hat{\epsilon}_{\pm}\right) = \sum_{m=0}^{k-1} \frac{1}{m!} y^m f^{(m)}(x)\,{}_k\hat{\epsilon}_{\pm}^{m}\,. \tag{7}$$

For example, setting $k = 2$, reproduces the well-known dual number identity [19]

$$f\left(x\,{}_2\hat{1} + y\,{}_2\hat{\epsilon}_{+}\right) = f(x)\,{}_2\hat{1} + yf'(x)\,{}_2\hat{\epsilon}_{+} = \begin{pmatrix} f(x) & yf'(x) \\ 0 & f(x) \end{pmatrix}. \tag{8}$$

For illustration, setting $k = 3$, we obtain

$$f\left(x\,{}_3\hat{1} + y\,{}_3\hat{\epsilon}_+\right) = f(x)\,{}_3\hat{1} + yf'(x)\,{}_3\hat{\epsilon}_+ + \tfrac{1}{2}y^2 f''(x)\,{}_3\hat{\epsilon}_+^2 = \begin{pmatrix} f(x) & yf'(x) & \frac{1}{2}y^2 f''(x) \\ 0 & f(x) & yf'(x) \\ 0 & 0 & f(x) \end{pmatrix}. \tag{9}$$

It may be verified that, e.g., for the choice "+" in Equation (5), the first row of the resulting matrices in Equation (7) contain the terms of the Taylor expansion up to order $k-1$. More explicitly, introducing the auxiliary notations for the row vector $\langle {}_k e_1|$ and the column vector $|{}_k\mathbf{1}\rangle$ of length $k \geq 2$,

$$\langle {}_k e_1| := (1, 0, \ldots, 0)\,, \quad |{}_k\mathbf{1}\rangle := \begin{pmatrix} 1 \\ \vdots \\ 1 \end{pmatrix}, \tag{10}$$

allows us to define the *order k evaluation operation* acting on some function $F({}_k\hat{\epsilon}_+)$ (which is itself assumed to act entry-wise) depending on a generalized dual number ${}_k\hat{\epsilon}_+$ as

$$\langle F({}_k\hat{\epsilon}_+)\rangle_k := \langle {}_k e_1|\, F({}_k\hat{\epsilon}_+)\, |{}_k\mathbf{1}\rangle\,. \tag{11}$$

We thus find that

$$\langle f\left(x\,{}_k\hat{1} + y\,{}_k\hat{\epsilon}_+\right)\rangle_k = \sum_{m=0}^{k-1} \frac{y^m}{m!} f^{(m)}(x)\,. \tag{12}$$

Recently, expansions such as Equation (7) have received considerable interest in the field of numerical analysis [20]. Referring to Fike [19] for an overview, various alternative types of "numbers" have been studied for the purpose of finding optimized numerical schemes for computing kth-order derivatives of functions. For example, it has been demonstrated that the use of so-called *hyper-dual numbers* results in first- and second-derivative calculations that are exact, regardless of the step size [14].

For later convenience, motivated by the identity (for $k \geq 2$)

$$\exp\left({}_k\hat{\epsilon}_+\, x\right) = \sum_{r=0}^{k-1} \frac{x^r}{r!}\, {}_k\hat{\epsilon}_+^r\,, \tag{13}$$

we may introduce the so-called *truncated exponential polynomials* [21] $e_n(x)$ defined through the series

$$e_n(x) := \sum_{r=0}^{n} \frac{x^r}{r!}\,, \tag{14}$$

which may be expressed in terms of generalized dual numbers as

$$\begin{pmatrix} e_n(x) \\ e_{n-1}(x) \\ \vdots \\ e_1(x) \\ 1 \end{pmatrix} := \exp\left({}_{n+1}\hat{\epsilon}_+\, x\right) |{}_{n+1}\mathbf{1}\rangle\,. \tag{15}$$

One may thus easily verify (via Equations (13) and (14)) the property

$$e_n'(x) = e_{n-1}(x)\,. \tag{16}$$

Having provided a matrix-based extension of ordinary to kth-order dual numbers of arbitrary order $k \geq 2$, we now proceed to develop a computational procedure embedding dual numbers with other techniques inspired by the operational umbral formalism.

3. Umbral-Type Methods and Dual Numbers

Starting from this section, we employ the notational simplification of writing ϵ for the *dual number unit (DNU)* ${}_k\hat{e}_\pm$ of generalized dual numbers (cf. Equation (5)), making the order $k \geq 2$ of the DN explicit only via the analog of the notation in Equation (11), and masking the matrix nature of ${}_k\hat{e}_\pm$. Thus, for some function $F \equiv F(\epsilon)$, we write

$$F \rightsquigarrow_k G \quad :\Leftrightarrow \quad G = \left(F\big|_{\epsilon^{k+1}\to 0}\right)\big|_{\epsilon\to 1} \tag{17}$$

for the truncation of F via setting $\epsilon^{k+1} = 0$ and afterwards $\epsilon = 1$. It is straightforward to verify that this formal definition may be implemented in terms of the matrix representations introduced in Section 2 via use of Equation (11) as

$$G = \langle F({}_{k+1}\hat{e}_+)\rangle_{k+1}\,. \tag{18}$$

Consider then the *dual complex parameter*

$$\hat{z} \equiv \hat{z}(a,b) := a + \epsilon b\,. \tag{19}$$

Following the principles of umbral calculus, we treat the dual complex parameter $\hat{z}$ as an ordinary algebraic quantity in calculations of integrals, derivatives and other operations, delaying the evaluation of $\hat{z}$ via performing the operation $\rightsquigarrow_k$ to the very end of the computations. Note that albeit the term umbral calculus has been introduced in the seminal papers by Roman and Rota [22], in the following we make reference to the formalism developed in [23] which enriches the original formalism with the wealth of techniques derived from the operational calculus [22,23]. We now illustrate the computational benefits of this approach via a number of examples.

3.1. Dual Shifted Gaussians

We first consider a Gaussian-type function explicitly containing in its argument the dual complex parameter in Equation (19), whence the *dual-shifted Gaussian* function

$$f(x) = e^{-\alpha x^2 + \hat{z}(a,b)\,x}\,. \tag{20}$$

Assuming for instance third-order dual numbers (i.e., $\epsilon^3 = 0$), we may write the above function in more conventional terms as

$$f(x) \rightsquigarrow_2 e^{-\alpha x^2 + ax}\left[1 + bx + \tfrac{1}{2}(bx)^2\right]\,, \tag{21}$$

which is easily recognized as the product of a shifted Gaussian with a second-degree polynomial.

In full analogy to the umbral operational methods of Licciardi [23], it is then straightforward to calculate the following integral of the function f of Equation (20) via the standard Gaussian integral formula

$$\int_{-\infty}^{+\infty} f(x)dx = \sqrt{\frac{\pi}{\alpha}}\, e^{\frac{\hat{z}(a,b)^2}{4\alpha}} = \sqrt{\frac{\pi}{\alpha}}\, e^{\frac{a^2}{4\alpha} + \frac{ab}{2\alpha}\epsilon + \frac{b^2}{4\alpha}\epsilon^2}\,. \tag{22}$$

The term on the right has in fact a definite meaning, since the use of the generating function of the *two-variable Hermite polynomials* [24]

$$\sum_{n=0}^{\infty} \frac{t^n}{n!} H_n(x,y) = e^{xt+yt^2} \tag{23a}$$

$$H_n(x,y) = e^{y\partial_x^2} x^n = n! \sum_{r=0}^{\lfloor \frac{n}{2} \rfloor} \frac{x^{n-2r} y^r}{(n-2r)!r!} \tag{23b}$$

permits to cast the right-hand side of Equation (22) into the form

$$\sqrt{\frac{\pi}{\alpha}}\, e^{\frac{a^2}{4\alpha}+\frac{ab}{2\alpha}\epsilon+\frac{b^2}{4\alpha}\epsilon^2} = \sqrt{\frac{\pi}{\alpha}}\, e^{\frac{a^2}{4\alpha}} \sum_{m\geq 0} \frac{\epsilon^m}{m!} H_m\left(\frac{ab}{2\alpha}, \frac{b^2}{4\alpha}\right) \rightsquigarrow_k \sqrt{\frac{\pi}{\alpha}}\, e^{\frac{a^2}{4\alpha}}\, {}_He_k\left(\frac{ab}{2\alpha}, \frac{b^2}{4\alpha}\right). \tag{24}$$

Here, ${}_He_k(x,y)$ denotes the *Hermite-based truncated exponential polynomial* [25–27] defined as

$${}_He_k(x,y) := \sum_{r=0}^{k} \frac{1}{r!} H_r(x,y). \tag{25}$$

3.2. Another Form of Dual Gaussian

Let us consider as a further example

$$g(x) := e^{-\hat{z}(a,b)x^2} \tag{26}$$

and the following infinite integral (for $\mathrm{Re}(a) > 0$)

$$\int_{-\infty}^{+\infty} g(x)dx = \sqrt{\frac{\pi}{\hat{z}}} = \sqrt{\frac{\pi}{a+\epsilon b}} \rightsquigarrow_k \sqrt{\frac{\pi}{a}} \sum_{r=0}^{k} \binom{-\frac{1}{2}}{r} \left(\frac{b}{a}\right)^r. \tag{27}$$

Here, by invoking the operation $\rightsquigarrow_k$, we obtain a finite series, thus obviating the need to impose any condition on the relevant convergence range.

3.3. Examples From Symbolic Calculus

The calculus of higher-order dual numbers may be further refined via combining it with the wealth of techniques available from the theory of special functions and symbolic calculus as put forward in [23,28–32]. Consider for illustration the following identity, known from the theory of two-variable Hermite polynomials [33],

$$\partial_x^n e^{\alpha x^2} = H_n(2\alpha x, \alpha) e^{\alpha x^2} \tag{28}$$

which allows to simplify the task of calculating successive derivatives of the dual Gaussian introduced in Equation (26), such as in the computation

$$\partial_x^n e^{-\hat{z}x^2} = H_n(-2\hat{z}x, -\hat{z}) e^{-\hat{z}x^2} = n! \sum_{r=0}^{\lfloor \frac{n}{2} \rfloor} \sum_{s\geq 0} \frac{(-1)^{n-r+s} 2^{n-2r} x^{n-2(r-s)}}{(n-2r)!r!s!} \hat{z}^{n-r+s}.$$

Here, the first step follows from (28) and the second by invoking (23b).

Another interesting type of calculus concerns infinite integrals involving rational functions such as

$$\Phi(x;a,b) := \frac{1}{1+\hat{z}x^2} \rightsquigarrow_k \frac{1}{1+ax^2} \sum_{r=0}^{k} \left(-\frac{bx^2}{1+ax^2}\right)^r. \tag{29}$$

For example, the infinite integral

$$\int_{-\infty}^{+\infty} \frac{1}{1+\hat{z}x^2} dx = \frac{\pi}{\sqrt{\hat{z}}} \tag{30}$$

may be easily transformed into truncated form in full analogy to the calculation summarized in Equation (27).

3.4. Umbral Image Type Techniques

Referring to Behr et al. [34] for the precise technical details (compare also [33]), suffice it here to provide the following definition for the action of the *formal integration operator* $\hat{\mathbb{I}}$ on the formal variable v (for $\alpha \in \mathbb{C}$):

$$\hat{\mathbb{I}}(v^\alpha) := \frac{1}{\Gamma(\alpha)} . \tag{31}$$

Then, an interesting variant of the example presented in Equation (29) may be obtained as

$$\hat{\mathbb{I}}\left[\int_{-\infty}^{+\infty} v\Phi(x;a,vb)dx\right] = \hat{\mathbb{I}}\left[\frac{v\pi}{\sqrt{z(a,v\beta)}}\right] \rightsquigarrow_k \sqrt{\frac{\pi}{a}} \sum_{r=0}^{k} \frac{1}{\Gamma(\frac{1}{2}-r)(r!)^2}\left(\frac{b}{a}\right)^r . \tag{32}$$

In summary, the combination of the concept of higher order dual numbers with techniques from symbolic and umbral-image type calculus appears to offer a large potential in view of novel tools of computation. To corroborate this claim, we now present some first high-level results in this direction.

4. Dual Numbers and Solution of Heat- and Schrödinger-Type Equations

Before entering the main topic of this section, let us recall a few useful "operational rules", starting with the *Glaisher identity* [27,35]

$$e^{\tau \frac{d^2}{dx^2}} e^{-\alpha x^2} = \frac{1}{\sqrt{1+4\tau\alpha}} e^{-\frac{\alpha x^2}{1+4\tau\alpha}} , \tag{33}$$

which can also be understood as the solution of the heat equation with a Gaussian as initial function. It proves particularly useful in the following to note that, according to the definition of the Hermite polynomials $H_n(x,y)$ as given in Equation (23b), an alternative interpretation of Equation (33) is provided in terms of the *double-lacunary exponential generating function* $\mathcal{H}_{2,0}(\lambda;x,y)$ of the polynomials $H_n(x,y)$, where we employ notations as in [36]

$$e^{\tau \frac{d^2}{dx^2}} e^{-\alpha x^2} = \sum_{n\geq 0} \frac{(-\alpha)^n}{n!} H_{2n}(x,\tau) = \mathcal{H}_{2,0}(-\alpha;x,\tau) . \tag{34}$$

By specializing Equation (33) to $\alpha = \hat{z}$ (with $\hat{z} = a + \epsilon b$ the dual complex parameter in Equation (19)), we obtain the operational identity

$$e^{\tau \frac{d^2}{dx^2}} e^{-\hat{z} x^2} = \frac{1}{\sqrt{1+4\tau\hat{z}}} e^{-\frac{\hat{z} x^2}{1+4\tau\hat{z}}} . \tag{35}$$

Via the simple factorizations

$$\begin{aligned} 1+4\hat{z}\tau &= \gamma(a,\tau)\gamma\left(\tfrac{b\epsilon}{\gamma(a,\tau)},\tau\right), \\ \frac{\hat{z}}{1+4\hat{z}\tau} &= \frac{a}{\gamma(a,\tau)} + \frac{b\epsilon}{[\gamma(a,\tau)]^2\gamma\left(\frac{b\epsilon}{\gamma(a,\tau)},\tau\right)}, \\ \gamma(c,\tau) &= 1+4c\tau , \end{aligned} \tag{36}$$

we may transform the identity in Equation (35) as

$$e^{\tau \frac{d^2}{dx^2}} e^{-\hat{z}x^2} = \mathcal{H}_{2,0}\left(-\frac{b\epsilon}{[\gamma(a,\tau)]^2}; x, \tau\gamma(a,\tau)\right) \mathcal{H}_{2,0}(-a; x, \tau) . \tag{37}$$

By re-inserting the definition of the first double-lacunary exponential generating function, using the Glaisher-identity in Equation (33) for the second one and finally truncating to order k, we eventually arrive at the compact result

$$e^{\tau \frac{d^2}{dx^2}} e^{-\hat{z}x^2} \rightsquigarrow_k \frac{e^{-\frac{ax^2}{\gamma(a,\tau)}}}{\sqrt{\gamma(a,\tau)}} \sum_{n=0}^{k} \frac{1}{n!} \left(\frac{b}{[\gamma(a,\tau)]^2}\right)^n H_{2n}\left(x, \tau\gamma(a,\tau)\right) , \qquad \gamma(a,\tau) = 1 + 4a\tau . \tag{38}$$

For example, by evaluating the above expression for second order dual numbers, one finds

$$e^{\tau \frac{d^2}{dx^2}} e^{-\hat{z}x^2} \rightsquigarrow_2 \frac{e^{-\frac{ax^2}{\gamma(a,\tau)}}}{\sqrt{\gamma(a,\tau)}} \left(1 - \frac{b}{\gamma(a,\tau)^2} H_2(x, \tau\gamma(a,\tau)) + \frac{b^2}{2\gamma(a,\tau)^4} H_4(x, \tau\gamma(a,\tau))\right) . \tag{39}$$

The above result may be interpreted as the solution of the heat-type equation

$$\partial_\tau F(x,\tau) = \partial_x^2 F(x,\tau) , \quad F(x,0) = e^{-\hat{z}x^2} . \tag{40}$$

An analogous problem has been addressed in [33] within the framework of a different method. The techniques we have envisaged may be further exploited to treat the *paraxial propagation* of the so-called *flattened distributions*, introduced in [37] to study the laser field evolution in optical cavities employing super-Gaussian mirrors [38]. These cavities shape beams whose transverse distribution is not reproduced by a simple Gaussian, but by a function exhibiting a *quasi-constant flat-top*, expressible through a function of the type

$$E(x;p) := e^{-|x|^p} , \qquad p \in \mathbb{Z}_{>0} . \tag{41}$$

The paraxial propagation of these beams has less obvious properties than, e.g., Laguerre or Hermite Gauss modes [38]. To overcome this drawback, Gori introduced the so-called *flattened beams* [37], which permit a fairly natural expansion in terms of Gauss Laguerre/Hermite modes, thus providing a straightforward solution to the corresponding paraxial wave equation.

Invoking our formalism as developed thus far, we may approximate the aforementioned Gori beams in the form

$$E(x;p) \approx Y(x;\alpha|m) := e^{-\alpha x^2} e_m(x^2) . \tag{42}$$

Here, $e_m(x)$ denotes the truncated exponential polynomials introduced in Equation (14), and both parameters α and m depend on p (see [33] for further details). Recalling from Equation (19) the definition $\hat{z}(a,b) := a + b\epsilon$ of the dual complex parameter, the r.h.s. of Equation (42) may be equivalently expressed as

$$e^{-\hat{z}(\alpha,-1)x^2} \rightsquigarrow_m Y(x;\alpha|m) , \tag{43}$$

whence as an instance of a *dual Gaussian* as described in Section 3.2. The problem of the relevant propagation can accordingly be reduced to that of an ordinary Gaussian mode, namely to the solution of the Schrödinger type equation

$$i\partial_\tau \Psi(x,\tau) = -\partial_x^2 \Psi(x,\tau) , \quad \Psi(x,0) = Y(x;\alpha|m) . \tag{44}$$

Consequently, by invoking the operational identity in Equation (35), the paraxial evolution of a flattened beam may be expressed in the form

$$\Psi(x,\tau) = e^{i\tau\partial_x^2} e^{-\hat{z}(\alpha,-1)x^2} = \frac{1}{\sqrt{1+4i\tau\hat{z}(\alpha,-1)}} e^{-\frac{\hat{z}(\alpha,-1)x^2}{1+4i\tau\hat{z}(\alpha,-1)}}, \tag{45}$$

which reproduces indeed the known solution of our problem (compare [33]).

In a forthcoming paper, we will discuss this specific application in further detail by applying the method to the problem of designing super-Gaussian optical systems.

5. Weyl Formula and Modified Hermite Polynomials

The wide flexibility of the method we propose is corroborated by the following further example, relevant to the use of operational ordering tools. Let us consider an evolution equation of the form

$$\partial_\tau F(x,\tau) = [\gamma\partial_x - \hat{z}x]F(x,\tau), \quad F(x,0) = f(x). \tag{46}$$

The relevant procedure for combining differential calculus with the umbral formalism is described in [39]. Following this approach, the solution of Equation (46) can be expressed as

$$F(x,\tau) = e^{\tau(\gamma\partial_x - \hat{z}x)} f(x). \tag{47}$$

To evaluate the solution of Equation (47) explicitly, we need to suitably "factorize" the exponential operator. This so-called *disentanglement operation* may be implemented via the Weyl formula [40]

$$e^{\hat{X}+\hat{Y}} = e^{-\frac{1}{2}[\hat{X},\hat{Y}]} e^{\hat{X}} e^{\hat{Y}}, \tag{48}$$

which is applicable whenever the identities $[\hat{X},[\hat{X},\hat{Y}]] = [[\hat{X},\hat{Y}],\hat{Y}] = 0$ hold. Applying the Weyl formula in Equation (48) to our solution in Equation (47), if we let $\hat{X} = -\tau\hat{z}x$ and $\hat{Y} = \tau\gamma\partial_x$ (resulting in $[\hat{X},\hat{Y}] = \tau^2\gamma\hat{z}$, and with higher-order commutators vanishing), we obtain

$$F(x,\tau) = e^{-\frac{1}{2}\tau^2\gamma\hat{z}} e^{-\hat{z}x\tau} f(x+\gamma\tau). \tag{49}$$

Thus, the solution at any desired truncation order k may be obtained by invoking the dual number evaluation operation $\rightsquigarrow_k$ of Equation (17).

As already mentioned above, the Weyl formula applies in the example presented because the algebraic structure of the argument of the exponential in Equation (47) satisfies a special property: the commutators of the associated generators reduce to a constant after the first commutation bracket. A more interesting extension is given by the case in which the generators are embedded into a solvable Lie algebra. In this case, the combined use of the dual number formalism and of the Wei-Norman ordering method [41] leads to new and interesting results. They deserve a separate treatment that will be reported in a forthcoming paper.

As a final example, we define *modified Hermite polynomials* $H_n(x,\hat{z})$, whence ordinary two-variable Hermite polynomials $H_n(x,y)$ as introduced in Equation (23b) evaluated at $y = \hat{z}$, with $\hat{z} \equiv \hat{z}(a,b)$ the dual complex parameter of Equation (19),

$$H_n(x,\hat{z}) = e^{\hat{z}\partial_x^2} x^n. \tag{50}$$

It is straightforward to verify that these modified polynomials inherit all the relevant properties from the polynomials $H_n(x,y)$, such as the recurrences

$$\begin{aligned} \partial_x H_n(x,\hat{z}) &= nH_{n-1}(x,\hat{z}), \\ H_{n+1}(x,\hat{z}) &= xH_n(x,\hat{z}) + 2\hat{z}\partial_x H_n(x,\hat{z}), \end{aligned} \tag{51}$$

and we find that they satisfy the second order differential equation

$$2\hat{z}\partial_x^2 H_n(x,\hat{z}) + x\partial_x H_n(x,\hat{z}) = nH_n(x,\hat{z}) \,. \tag{52}$$

The explicit form of these truncated polynomials is easily obtained. For example, by using third order dual numbers, which implies

$$e^{\hat{z}\partial_x^2} \rightsquigarrow_2 e^{a\partial_x^2}\left(1 + b\partial_x^2 + \frac{1}{2}b^2\partial_x^4\right), \tag{53}$$

we find the explicit formula

$$H_n(x,\hat{z}) \rightsquigarrow_2 H_n(x,a) + b\partial_a H_n(x,a) + \tfrac{1}{2}b^2\partial_a^2 H_n(x,a)\,, \tag{54}$$

where we have invoked the well-known identity

$$\partial_x^2 H_n(x,y) = \partial_y H_n(x,y)\,. \tag{55}$$

6. Final Comments

The method we have outlined in this paper offers many computational advantages to treat problems where truncated expansions (not necessarily of Taylor type) of functions are involved. At its core, the umbral formalism and the notion of higher order dual numbers allow delaying the explicit expansions to later stages in a given calculation, thus opening the possibility to exploit numerous efficient computation strategies from the theory of operational calculus and special functions.

The technique we have introduced in this paper is amenable for new applications in various different fields. We have presented herein the solution of parabolic equations in transport problems, and within such a context a fairly important example has been provided by treating the propagation of flattened beams [33,37] in optics. For brevity, we have just outlined the procedure in terms of a one-dimensional computation. The relevant extension to the three-dimensional case does not require any particular conceptual effort, but only a consistent numerical implementation. In a forthcoming investigation, we will further extend the method and study its potential for treating perturbative problems in classical and quantum mechanics.

Author Contributions: Conceptualization, G.D.; methodology, N.B. and G.D.; validation, N.B., G.D. and S.L.; formal analysis, N.B., G.D., A.L., and S.L.; writing—original draft preparation, N.B., G.D., and A.L.; and writing—review and editing, N.B., A.L., and S.L.

Funding: The work of N.B. is supported by funding from the European Union's Horizon 2020 research and innovation programme under the Marie Skłodowska-Curie grant agreement No. 753750. A.L. was supported by the NCN research project OPUS 12 No. UMO-2016/23/B/ST3/01714 and by the NAWA project: Program im. Iwanowskiej PPN/IWA/2018/1/00098. S.L. was supported by an *Enea-Research Center Individual Fellowship*.

Acknowledgments: N.B. would like to thank the LPTMC (Paris 06) and ENEA Frascati for warm hospitality.

Conflicts of Interest: The authors declare no conflict of interest. The founding sponsors had no role in the design of the study; in the collection, analyses, or interpretation of data; in the writing of the manuscript, and in the decision to publish the results.

References

1. Clifford, W.K. Preliminary Sketch of Biquaternions. *Proc. Lond. Math. Soc.* **1871**, *s1–s4*, 381–395. [CrossRef]
2. Grünwald, J. Über duale Zahlen und ihre Anwendung in der Geometrie. *Monatshefte für Mathematik Physik* **1906**, *17*, 81–136. [CrossRef]
3. Segre, C. *Le Geometrie Projettive nei Campi di Numeri Duali*; Vincenzo Bona: Torino, Italy, 1912.
4. Yaglom, I.M. *A Simple Non-Euclidean Geometry and Its Physical Basis: An Elementary Account of Galilean Geometry and the Galilean Principle of Relativity*; Springer: Berlin, Germany, 2012.

5. Klawitter, D. *Clifford Algebras: Geometric Modelling and Chain Geometries with Application in Kinematics*; Springer: Berlin, Germany, 2014.
6. Kotelnikov, A.P. Screw Calculus and Some Applications to Geometry and Mechanics. *Ann. Imp. Univ. Kazan* **1895**, *24*. (In Russian)
7. Study, E. Die Geometrie der Dynamen. *Jahresbericht der Deutschen Mathematiker-Vereinigung* **1900**, *8*, 204–216.
8. Rooney, J. On the principle of transference. In *Proceedings of the Fourth World Congress on the Theory of Machines and Mechanisms*; Institution of Mechanical Engineers: London, UK, 1975; pp. 1088–1092.
9. Hsia, L.; Yang, A. On the principle of transference in three-dimensional kinematics. *J. Mech. Des.* **1981**, *103*, 652–656. [CrossRef]
10. Martínez, J.M.R.; Duffy, J. The principle of transference: History, statement and proof. *Mech. Mach. Theory* **1993**, *28*, 165–177. [CrossRef]
11. Cohen, A.; Shoham, M. Application of Hyper-Dual Numbers to Multibody Kinematics. *J. Mech. Rob.* **2015**, *8*, 011015. [CrossRef]
12. Cohen, A.; Shoham, M. Application of hyper-dual numbers to rigid bodies equations of motion. *Mech. Mach. Theory* **2017**, *111*, 76–84. [CrossRef]
13. Cohen, A.; Shoham, M. Principle of transference—An extension to hyper-dual numbers. *Mech. Mach. Theory* **2018**, *125*, 101–110. [CrossRef]
14. Fike, J.; Alonso, J. The Development of Hyper-Dual Numbers for Exact Second-Derivative Calculations. In *Proceedings of the 49th AIAA Aerospace Sciences Meeting including the New Horizons Forum and Aerospace Exposition*; American Institute of Aeronautics and Astronautics: Reston, VA, USA, 2011. [CrossRef]
15. Harkin, A.A.; Harkin, J.B. Geometry of Generalized Complex Numbers. *Math. Mag.* **2004**, *77*, 118–129. [CrossRef]
16. Özdemir, M. Introduction to Hybrid Numbers. *Adv. Appl. Clifford Algebr.* **2018**, *28*. [CrossRef]
17. Rall, L.B. (Ed.) *Automatic Differentiation: Techniques and Applications*; Springer: Berlin/Heidelberg, Germany, 1981. [CrossRef]
18. Rall, L.B.; Corliss, G.F. An introduction to automatic differentiation. In *Computational Differentiation: Techniques, Applications, and Tools*; SIAM: Philadelphia, PA, USA, 1996, Volume 89.
19. Fike, J.A. *Derivative Calculations Using Hyper-Dual Numbers*; Technical Report; Sandia National Lab. (SNL-NM): Albuquerque, NM, USA, 2016.
20. Berland, H. (Department of Mathematical Sciences, NTNU, Trondheim, Norway). Personal communication, 2006.
21. Dattoli, G.; Cesarano, C.; Sacchetti, D. A note on truncated polynomials. *Appl. Math. Comput.* **2003**, *134*, 595–605. [CrossRef]
22. Roman, S.M.; Rota, G.C. The umbral calculus. *Adv. Math.* **1978**, *27*, 95–188. [CrossRef]
23. Licciardi, S. Umbral Calculus, a Different Mathematical Language. *arXiv* **2018**, arXiv:1803.03108.
24. Appel, P.; De Feriet, J.K. Fonctions hypergéométriques et hypersphériques. In *Polynômes d'Hermite*; Gauthier-Villars: Paris, France, 1926.
25. Dattoli, G. Hermite-Bessel and Laguerre-Bessel functions: A by-product of the monomiality principle. In *Proceedings of the Melfi School on Advanced Topics in Mathematics and Physics, Advanced Special Functions and Applications*; Aracne Editrice: Melfi, Italy, 2000; pp. 147–164.
26. Dattoli, G. Generalized polynomials, operational identities and their applications. *J. Comput. Appl. Math.* **2000**, *118*, 111–123. [CrossRef]
27. Dattoli, G.; Khan, S.; Ricci, P. On Crofton—Glaisher type relations and derivation of generating functions for Hermite polynomials including the multi-index case. *Integral Transf. Spec. Funct.* **2008**, *19*, 1–9. [CrossRef]
28. Babusci, D.; Dattoli, G.; Górska, K.; Penson, K. Repeated derivatives of composite functions and generalizations of the Leibniz rule. *Appl. Math. Comput.* **2014**, *241*, 193–199. [CrossRef]
29. Babusci, D.; Dattoli, G. On Ramanujan Master Theorem. *arXiv* **2011**, arXiv:1103.3947.
30. Górska, K.; Babusci, D.; Dattoli, G.; Duchamp, G.; Penson, K. The Ramanujan master theorem and its implications for special functions. *Appl. Math. Comput.* **2012**, *218*, 11466–11471. [CrossRef]
31. Dattoli, G.; Di Palma, E.; Sabia, E.; Gorska, K.; Horzela, A.; Penson, K. Operational versus umbral methods and the Borel transform. *Int. J. Appl. Comput. Math.* **2017**, *3*, 3489–3510. [CrossRef]
32. Babusci, D.; Dattoli, G.; Górska, K.; Penson, K. Lacunary generating functions for the Laguerre polynomials. *Séminaire Lotharingien Combinatoire* **2017**, *76*, B76b.

33. Dattoli, G.; Migliorati, M. The truncated exponential polynomials, the associated Hermite forms and applications. *Int. J. Math. Math. Sci.* **2006**, *2006*, 98175-1. [CrossRef]
34. Behr, N.; Dattoli, G.; Duchamp, G.; Licciardi, S.; Penson, K. Operational Methods in the Study of Sobolev-Jacobi Polynomials. *Mathematics* **2019**, *7*, 124. [CrossRef]
35. Crofton, M. On operative symbols in the differential calculus. *Proc. Lond. Math. Soc.* **1880**, *1*, 122–134. [CrossRef]
36. Behr, N.; Duchamp, G.H.; Penson, K.A. Explicit formulae for all higher order exponential lacunary generating functions of hermite polynomials. *arXiv* **2018**, arXiv:1806.08417.
37. Gori, F. Flattened gaussian beams. *Opt. Commun.* **1994**, *107*, 335–341. [CrossRef]
38. Siegman, A.E. Lasers university science books. *Mill Valley CA* **1986**, *37*, 208.
39. Babusci, D.; Dattoli, G. Umbral methods and operator ordering. *arXiv* **2011**, arXiv:1112.1570.
40. Dattoli, G.; Ottaviani, P.L.; Torre, A.; Vázquez, L. Evolution operator equations: Integration with algebraic and finite difference methods. Applications to physical problems in classical and quantum mechanics and quantum field theory. *La Rivista del Nuovo Cimento* **1997**, *20*, 3–133. [CrossRef]
41. Wei, J.; Norman, E. Lie algebraic solution of linear differential equations. *J. Math. Phys.* **1963**, *4*, 575–581. [CrossRef]

axioms

MDPI

Article

Branching Functions for Admissible Representations of Affine Lie Algebras and Super-Virasoro Algebras

Namhee Kwon

Department of Mathematics, Daegu University, Gyeongsan, Gyeongbuk 38453, Korea; nkwon@daegu.ac.kr

Received: 2 May 2019; Accepted: 17 July 2019; Published: 19 July 2019

Abstract: We explicitly calculate the branching functions arising from the tensor product decompositions between level 2 and principal admissible representations over $\widehat{\mathfrak{sl}}_2$. In addition, investigating the characters of the minimal series representations of super-Virasoro algebras, we present the tensor product decompositions in terms of the minimal series representations of super-Virasoro algebras for the case of principal admissible weights.

Keywords: branching functions; admissible representations; characters; affine Lie algebras; super-Virasoro algebras

1. Introduction

One of the basic problems in representation theory is to find the decomposition of a tensor product between two irreducible representations. In fact, the study of tensor product decompositions plays an important role in quantum mechanics and in string theory [1,2], and it has attracted much attention from combinatorial representation theory [3]. In addition, recent studies reveal that tensor product decompositions are also closely related to the representation theory of Virasoro algebra and W-algebras [4–6].

In [6], the authors extensively study decompositions of tensor products between integrable representations over affine Lie algebras. They also investigate relationships among tensor products, branching functions and Virasoro algebra through integrable representations over affine Lie algebras.

In the present paper we shall follow the methodology appearing in [6]. However, we will focus on admissible representations of affine Lie algebras. Admissible representations are not generally integrable over affine Lie algebras, but integrable with respect to a subroot system of the root system attached to a given affine Lie algebra. Kac and Wakimoto showed that admissible representations satisfy several nice properties such as Weyl-Kac type character formula and modular invariance [5,7]. In their subsequent works, they also established connections between admissible representations of affine Lie algebras and the representation theory of W-algebras [4,8]. In addition, Kac and Wakimoto expressed in ([9], Theorem 3.1) the branching functions arising from the tensor product decompositions between principal admissible and integrable representations as the q-series involving the associated dominant integral weights and string functions.

One of the main results of this paper is the explicit calculations of the branching functions appearing in ([9], Theorem 3.1). We are particularly interested in the calculations of the branching functions obtained from certain tensor product decompositions of level 2 integrable and principal admissible representations over $\widehat{\mathfrak{sl}}_2$ (see Theorem 4). We shall see that these branching functions connect the representation theory of affine Lie algebras with the representation theory of super-Virasoro algebras.

We usually apply the theory of modular functions for calculations of string functions [10]. However, in the current work we shall not use the tools of modular functions for the calculations of the string functions appearing in ([9], Theorem 3.1). Instead, we shall use both the invariance properties of string functions under the action of affine Weyl group and the character formula whose summation is taken over maximal weights (see Theorem 5). It seems like that this approach provides a simpler way for computations of the string functions in our cases.

We would like to point out that in ([5], Corollary 3(c)) the authors expressed the branching functions in terms of theta functions. We shall show that our expressions for the branching functions appearing in Theorem 4 are actually same as those of ([5], Corollary 3(c)) through the investigations of the characters of the minimal series representations of super-Virasoro algebras. Comparing our calculations of the branching functions over $\widehat{\mathfrak{sl}}_2$ with the characters of the minimal series representations of super-Virasoro algebras, we also present the tensor product decompositions between level 2 integrable and principal admissible representations in terms of the minimal series representations of super-Virasoro algebras (see Theorem 6). This generalizes the decomposition formula appearing in ([6], Section 4.1(a)) to the case of principal admissible weights.

2. Preliminaries

Let $A = (a_{ij})_{1\leq i,j\leq n}$ be a symmetrizable generalized Cartan matrix and $\mathfrak{g}$ the Kac-Moody Lie algebra associated with A. Let $\mathfrak{h}$ be a Cartan subalgebra of $\mathfrak{g}$. Fix the set of simple roots $\Pi = \{\alpha_1, \cdots, \alpha_n\}$ of $\mathfrak{h}$ and simple coroots $\Pi^\vee = \{\alpha_1^\vee, \cdots, \alpha_n^\vee\}$ of $\mathfrak{h}^*$, respectively. Assume that Π and $\Pi^\vee$ satisfy the condition $\alpha_j(\alpha_i^\vee) = a_{ij}$. We denote by $(\ |\ )$ the non-degenerate invariant symmetric bilinear form on $\mathfrak{g}$. Write Δ, Δ_+ and Δ_- for the set of all roots, positive roots and negative roots of $\mathfrak{g}$, respectively. Put $\Delta^{re} = \{\alpha \in \Delta \mid (\alpha|\alpha) > 0\}$ and $\Delta^{im} = \{\alpha \in \Delta \mid (\alpha|\alpha) \leq 0\}$. For each $i = 1, \cdots, n$, we define the fundamental reflection r_{α_i} of $\mathfrak{h}^*$ by

$$r_{\alpha_i}(\lambda) = \lambda - \lambda(\alpha_i^\vee)\,\alpha_i \ (\lambda \in \mathfrak{h}^*).$$

The subgroup W of $GL(\mathfrak{h}^*)$ generated by all fundamental reflections is called the *Weyl group* of $\mathfrak{g}$.

Among symmetrizable Kac-Moody Lie algebras, the most important Lie algebras are affine Lie algebras whose associated Cartan matrices are called *affine Cartan matrices*. It is known that every affine Cartan matrix is a positive semidefinite of corank 1. Each affine Cartan matrix is in one-to-one correspondence with the affine Dynkin diagram of type $X_n^{(r)}$, where $X = A, B, C, D, E, F$ or G and $r = 1, 2$ or 3. The number r is called the *tier number* (see [11,12] for details). Given an affine Cartan matrix $A = (a_{ij})_{0\leq i,j\leq l}$, two $(l+1)$-tuples $(a_i^\vee)_{0\leq i\leq l}$ and $(a_i)_{0\leq i\leq l}$ of positive integers are uniquely determined by the conditions

1. $(a_0^\vee, a_1^\vee, \cdots, a_l^\vee)\,A = \mathbf{O}$,
2. $A\begin{pmatrix} a_0 \\ a_1 \\ \vdots \\ a_l \end{pmatrix} = \mathbf{O}$,
3. $gcd(a_0^\vee, a_1^\vee, \cdots, a_l^\vee) = gcd(a_0, a_1, \cdots, a_l) = 1$,

where $\mathbf{O}$ is the zero vector. We call $(a_i)_{0\leq i\leq l}$ (resp. $(a_i^\vee)_{0\leq i\leq l}$) the *label* (resp. *colabel*) of the affine matrix A. The corresponding positive integer $h = \sum_{i=0}^{l} a_i$ (resp. $h^\vee = \sum_{i=0}^{l} a_i^\vee$) is called the *Coxeter number* (resp. *dual Coxeter number*). Notice that the element $K = \sum_{i=0}^{l} a_i^\vee \alpha_i^\vee$ satisfies $\alpha_i(K) = 0$ for $0 \leq i \leq l$, and we call this element the *central element*. Through the non-degenerate bilinear form $(\ |\ )$ defined on $\mathfrak{g}$, the central element K corresponds to $\delta = \sum_{i=0}^{l} a_i\alpha_i$ in $\mathfrak{h}^*$.

Suppose that $\mathfrak{g}$ is the affine Lie algebra associated to an affine Cartan matrix $A = (a_{ij})_{0\leq i,j\leq l}$, and let $\mathfrak{h}$ be a Cartan subalgebra of $\mathfrak{g}$. The Cartan subalgebra $\mathfrak{h}$ is $(l+2)$-dimensional, and we can decompose $\mathfrak{h}$ and $\mathfrak{h}^*$ as follows:

$$\mathfrak{h} = \overline{\mathfrak{h}} \oplus \mathbb{C}K \oplus \mathbb{C}d,$$

$$\mathfrak{h}^* = \overline{\mathfrak{h}}^* \oplus \mathbb{C}\delta \oplus \mathbb{C}\Lambda_0,$$

where $\overline{\mathfrak{h}} = \sum_{i=1}^{l} \mathbb{C}\alpha_i^\vee$ and $\overline{\mathfrak{h}}^* = \sum_{i=1}^{l} \mathbb{C}\alpha_i$.

The lattice $Q = \sum_{i=0}^{l} \mathbb{Z}\alpha_i$ and $Q^\vee = \sum_{i=0}^{l} \mathbb{Z}\alpha_i^\vee$ are called the *root lattice* and *coroot lattice*, respectively. Set

$$M = \begin{cases} \overline{Q}^\vee & \text{if } r = 1 \text{ or } A = A_{2l}^{(2)}, \\ \overline{Q} & \text{if } r \geq 2 \text{ and } A \neq A_{2l}^{(2)}. \end{cases}$$

For an element $\alpha \in Q$, we define $t_\alpha \in GL(\mathfrak{h}^*)$ by

$$t_\alpha(\lambda) = \lambda + (\lambda|\delta)\,\alpha - \left\{ \frac{|\alpha|^2}{2}(\lambda|\delta) + (\lambda|\alpha) \right\} \delta \ (\lambda \in \mathfrak{h}^*).$$

We call t_α $(\alpha \in Q)$ the *translation operator*. It is known that the Weyl group W of the affine Lie algebra $\mathfrak{g}$ is also given by $\overline{W} \ltimes t_M$, where $\overline{W} = \langle r_{\alpha_i} | 1 \leq i \leq l \rangle$ and $t_M = \{t_\alpha | \alpha \in M\}$.

For a non-twisted affine Lie algebra (i.e., $r = 1$), recall that

$$\Delta^{im} = \{n\delta | n \in \mathbb{Z} - \{0\}\}$$

and

$$\Delta^{re} = \{n\delta + \alpha | n \in \mathbb{Z},\ \alpha \in \overline{\Delta}\},$$

where $\overline{\Delta}$ is the set of all roots of the finite-dimensional simple Lie algebra associated with the finite Cartan matrix $\overline{A} = (a_{ij})_{1 \leq i,j \leq l}$.

Set

$$\begin{aligned} P &= \{\lambda \in \mathfrak{h}^* | \lambda(\alpha_i^\vee) \in \mathbb{Z} \text{ for } 0 \leq i \leq l\}, \\ P^m &= \{\lambda \in P | \lambda(K) = m\}, \\ P_+ &= \{\lambda \in \mathfrak{h}^* | \lambda(\alpha_i^\vee) \in \mathbb{Z}_{\geq 0} \text{ for } 0 \leq i \leq l\}, \\ P_+^m &= P^m \cap P_+. \end{aligned}$$

An element in P (reps. P_+) is called an *integral weight* (resp. a *dominant integral weight*). Let ρ be the dominant integral weight defined by $\rho(\alpha_i^\vee) = 1$ for $0 \leq i \leq l$. The element ρ is called the *Weyl vector* of $\mathfrak{g}$. It is sometimes convenient to choose the Weyl vector satisfying the additional condition $\rho(d) = 0$, and we get $\rho = \overline{\rho} + h^\vee \Lambda_0$ in this case.

Define the fundamental weights $\Lambda_i \in \mathfrak{h}^*$ $(0 \leq i \leq l)$ by $\Lambda_i\left(\alpha_j^\vee\right) = \delta_{ij}$ $(0 \leq j \leq l)$ and $\Lambda_i(d) = 0$. Similarly, we define the fundamental coweights $\Lambda_i^\vee \in \mathfrak{h}$ $(0 \leq i \leq l)$ by $(\Lambda_i^\vee|\alpha_j) = \delta_{ij}$ $(0 \leq j \leq l)$ and $(\Lambda_i^\vee|d) = 0$. Let $\overline{\Lambda}_i$ and $\overline{\Lambda}_i^\vee$ be the restrictions of Λ_i and $\Lambda_i^\vee$ to $\overline{\mathfrak{h}}^*$ and $\overline{\mathfrak{h}}$, respectively. Put $\overline{P} = \sum_{i=1}^{l} \mathbb{Z}\overline{\Lambda}_i$ and $\overline{P}^\vee = \sum_{i=1}^{l} \mathbb{Z}\overline{\Lambda}_i^\vee$, and let us introduce a lattice

$$\widetilde{M} = \begin{cases} \overline{P}^\vee & \text{if } r = 1 \text{ or } A = A_{2l}^{(2)}, \\ \overline{P} & \text{if } r \geq 2 \text{ and } A \neq A_{2l}^{(2)}. \end{cases}$$

Then, the group $\widetilde{W} = \overline{W} \ltimes t_{\widetilde{M}}$ is called the *extended affine Weyl group* of $\mathfrak{g}$.

3. Branching functions for admissible weights

Let $\mathfrak{g}$ be the Kac-Moody Lie algebra associated to a symmetrizable generalized Cartan matrix A, and $\mathfrak{h}$ a Cartan subalgebra of $\mathfrak{g}$. An element $\lambda \in \mathfrak{h}^*$ satisfying conditions

1. $\langle \lambda + \rho, \alpha^\vee \rangle \in \mathbb{Q} - \mathbb{Z}_{\leq 0}$ for all $\alpha \in \Delta_+^{re} := \Delta^{re} \cap \Delta_+$,
2. $\mathbb{Q}$-span of $\{\alpha \in \Delta_+^{re} | \langle \lambda + \rho, \alpha^\vee \rangle \in \mathbb{Z}\} = \mathbb{Q}$-span of Δ_+^{re}

is called an *admissible weight*. When λ is an admissible weight, the corresponding irreducible highest weight $\mathfrak{g}$-module $L(\lambda)$ is called an *admissible* $\mathfrak{g}$*-module* or *admissible representation*. Write

$$\Delta_{\lambda}^{\vee re} = \left\{ \alpha^{\vee} | \alpha \in \Delta^{re} \text{ and } \left\langle \lambda + \rho, \alpha^{\vee} \right\rangle \in \mathbb{Z}_{\geq 1} \right\}.$$

Then, it is easy to see that $\Delta_{\lambda}^{\vee re}$ forms a subroot system of the coroot system $\Delta^{\vee}$. We denote by $\Pi_{\lambda}^{\vee}$ a base of $\Delta_{\lambda}^{\vee\, re}$, and put $W_{\lambda} = \left\langle r_{\alpha} | \alpha \in \Pi_{\lambda}^{\vee} \right\rangle$.

An admissible weight λ is called a *principal admissible weight* if $\Pi_{\lambda}^{\vee}$ is isomorphic to $\Pi^{\vee}$. In general, the level of a principal admissible weight is a rational number. In fact, it is known from [7] that a rational number $m = \frac{v}{u}$ $(u \in \mathbb{Z}_{\geq 1},\ v \in \mathbb{Z},\ gcd(u,v) = 1)$ is the level of principal admissible weights if and only if it satisfies

1. $gcd\left(u, r^{\vee}\right) = 1$,
2. $u\left(m + h^{\vee}\right) \geq h^{\vee}$,

where $r^{\vee}$ is the tier number of the transposed generalized Cartan matrix A^{t} and $h^{\vee}$ denotes the dual Coxeter number of $\mathfrak{g}$.

Henceforth, we assume that $\mathfrak{g}$ is an affine Lie algebra with a simple coroot system $\Pi^{\vee} = \{\alpha_0^{\vee}, \cdots, \alpha_l^{\vee}\}$.

Given $u \in \mathbb{Z}_{\geq 1}$, put $\gamma_0 = (u-1)c + \alpha_0^{\vee}$ and $\gamma_i = \alpha_i^{\vee}$ $(1 \leq i \leq l)$. Define $S_{(u)} = \{\gamma_i \mid 0 \leq i \leq l\}$. Then, $S_{(u)}$ becomes a simple coroot system of $\Delta^{\vee} \cap \left(\sum_{i=0}^{l} \mathbb{Z}\gamma_i\right)$ if $gcd\left(u, r^{\vee}\right) = 1$ (see [13], Lemma 3.2.1). Moreover, the following theorems are known.

Theorem 1. *Let* $m = \frac{v}{u}$ *with* $u \in \mathbb{Z}_{\geq 1}, v \in \mathbb{Z}$ *and* $gcd(u,v) = 1$. *Assume that* $y \in \widetilde{W}$ *satisfies* $y\left(S_{(u)}\right) \subset \Delta_{+}^{\vee}$. *Write* $P_{u,y}^{m}$ *for the set of all principal admissible weights* λ *of level* m *with* $\Pi_{\lambda}^{\vee} = y\left(S_{(u)}\right)$. *Then, we have*

$$P_{u,y}^{m} = \left\{ y\left(\lambda^0 - (u-1)\left(m + h^{\vee}\right)\Lambda_0 + \rho\right) - \rho \mid \lambda^0 \in P_{+}^{u(m+h^{\vee})-h^{\vee}} \right\}.$$

Proof. See ([7], Theorem 2.1) or ([9], Proposition 1.5). □

Theorem 2. *Let* $m = \frac{v}{u}$ *with* $u \in \mathbb{Z}_{\geq 1}, v \in \mathbb{Z}$ *and* $gcd(u,v) = 1$. *Let* P_{+}^{m} *be the set of all principal admissible weights of level* m *(we use the same notation as the case of dominant integral weights). Then,* $P_{+}^{m} = \bigcup_y P_{u,y}^{m}$, *where* y *runs over* $\left\{ y \in \widetilde{W} | y\left(S_{(u)}\right) \subset \Delta_{+}^{\vee} \right\}$.

Proof. See ([9], Proposition 1.5). □

Let us now review branching functions and their connections with the Virasoro algebra.

Recall the Virasoro algebra is an infinite dimensional Lie algebra $Vir = \left(\bigoplus_{n \in \mathbb{Z}} \mathbb{C}\ell_n\right) \oplus \mathbb{C}c$ with brackets

$$[\ell_m, c] = 0 \text{ for all } m \in \mathbb{Z}$$

and

$$[\ell_m, \ell_n] = (m-n)\ell_{m+n} + \frac{m^3 - m}{12}\delta_{m+n,0}\, c \text{ for all } m, n \in \mathbb{Z}.$$

Let $\bar{\mathfrak{g}}$ be a finite dimensional simple Lie algebra, and $\mathfrak{g} = \mathbb{C}\left[t, t^{-1}\right] \otimes \bar{\mathfrak{g}} \oplus \mathbb{C}K \oplus \mathbb{C}d$ the non-twisted affine Lie algebra over $\bar{\mathfrak{g}}$. Let V be the highest weight $\mathfrak{g}$-module of level m such that $m + h^{\vee} \neq 0$. Define the operators $L_n^{\mathfrak{g}}$ $(n \in \mathbb{Z})$ via

$$L^{\mathfrak{g}}(z) = \sum_{n \in \mathbb{Z}} L_n^{\mathfrak{g}} z^{-n-2} = \frac{1}{2\left(m + h^{\vee}\right)} \sum_{i=1}^{dim \bar{\mathfrak{g}}} : u^{i}(z) u_i(z) :, \tag{1}$$

where $\{u^i\}$ and $\{u_i\}$ are bases of $\overline{\mathfrak{g}}$ satisfying $(u_i|u^j) = \delta_{ij}$. It is well-known that V becomes a *Vir*-module by letting

$$\ell_n \longmapsto L_n^{\mathfrak{g}} \ (n \in \mathbb{Z}) \text{ and } c \longmapsto \frac{m\dim\overline{\mathfrak{g}}}{m+h^\vee}. \tag{2}$$

The Virasoro action (2) satisfies the following properties:

$$\left[\ell_n, t^j \otimes X\right] = -jt^{j+n} \otimes X \ (X \in \overline{\mathfrak{g}},\ n \in \mathbb{Z} - \{0\},\ j \in \mathbb{Z}), \tag{3}$$

$$\ell_0 = \frac{(\Lambda + 2\rho|\Lambda)}{2(m+h^\vee)}\mathrm{Id} - d. \tag{4}$$

Let $\overline{\mathfrak{p}}$ be a reductive subalgebra of $\overline{\mathfrak{g}}$. Then, $\overline{\mathfrak{p}}$ is decomposed as $\overline{\mathfrak{p}} = \overline{\mathfrak{p}}_0 \oplus \overline{\mathfrak{p}}_1 \oplus \cdots \oplus \overline{\mathfrak{p}}_s$, where $\overline{\mathfrak{p}}_0$ is the center of $\overline{\mathfrak{p}}$ and each $\overline{\mathfrak{p}}_i$ $(i = 1, \cdots, s)$ is a simple Lie algebra. Assume that

$$\overline{\mathfrak{p}}_0 \oplus \left(\sum_{i=1}^{s} \overline{\mathfrak{h}}_i\right) \subset \overline{\mathfrak{h}}$$

and

$$\sum_{i=1}^{s} \overline{\mathfrak{p}}_{i+} \subset \overline{\mathfrak{g}}_+,$$

where $\overline{\mathfrak{h}}_i$ (resp. $\overline{\mathfrak{h}}$) is a Cartan subalgebra of $\overline{\mathfrak{p}}_i$ (resp. $\overline{\mathfrak{g}}$) and $\overline{\mathfrak{p}}_{i+}$ (resp. $\overline{\mathfrak{g}}_+$) is the sum of the positive root spaces of $\overline{\mathfrak{p}}_i$ (resp. $\overline{\mathfrak{g}}$). Consider the affinization $\mathfrak{p} = (\mathbb{C}[t,t^{-1}] \otimes \overline{\mathfrak{p}}) \oplus \mathbb{C}\dot{K} \oplus \mathbb{C}d$ of $\overline{\mathfrak{p}}$. Since V is the highest weight $\mathfrak{g}$-module, V is also the highest weight $\mathfrak{p}$-module. However, the action of the central element $\dot{K}$ on V is somewhat complicated. We refer to ([11], Chapter 12) for the details of the action of the central element $\dot{K}$. Let $\dot{m}$ be the level of V as a $\mathfrak{p}$-module, and write $(\ |\ )'$ for the standard bilinear form on $\overline{\mathfrak{p}}$. Set

$$L^{\mathfrak{p}} = \sum_{n\in\mathbb{Z}} L_n^{\mathfrak{p}} z^{-n-2} = \frac{1}{2(\dot{m}+\dot{h}^\vee)} \sum_{i=1}^{\dim\overline{\mathfrak{p}}} :\dot{u}^i(z)\dot{u}_i(z):, \tag{5}$$

where $\{\dot{u}^i\}$ and $\{\dot{u}_i\}$ are bases of $\overline{\mathfrak{p}}$ satisfying $(\dot{u}^i|\dot{u}_j)' = \delta_{ij}$ and $\dot{h}^\vee$ is the dual Coxeter number of $\overline{\mathfrak{p}}$. Using (1) and (5), define

$$L^{\mathfrak{g};\mathfrak{p}}(z) = L^{\mathfrak{g}} - L^{\mathfrak{p}} = \sum_{n\in\mathbb{Z}} L_n^{\mathfrak{g};\mathfrak{p}} z^{-n-2}.$$

Due to (3), it follows that

$$\left[L_n^{\mathfrak{g};\mathfrak{p}}, t^j \otimes X\right] = 0 \text{ for all } X \in \overline{\mathfrak{p}},\ n \in \mathbb{Z} - \{0\} \text{ and } j \in \mathbb{Z}. \tag{6}$$

Applying the operator product expansions, we can verify that $L^{\mathfrak{g};\mathfrak{p}}(z)$ is, in fact, a Virasoro field with the central charge $c^{\mathfrak{g};\mathfrak{p}} = \frac{m\dim\overline{\mathfrak{g}}}{m+h^\vee} - \frac{\dot{m}\dim\overline{\mathfrak{p}}}{\dot{m}+\dot{h}^\vee}$ (see [13,14] for the details). We call the Virasoro field $L^{\mathfrak{g};\mathfrak{p}}(z)$ the *coset Virasoro field*.

In the remaining part of this section, we assume that $V = L(\Lambda)$ for a dominant integral weight Λ of level m. Let $\dot{\overline{\mathfrak{h}}}$ be a Cartan subalgebra of $\overline{\mathfrak{p}}$, and $\mathfrak{p}_+$ the positive part of $\mathfrak{p}$. For $\nu \in \left(\dot{\overline{\mathfrak{h}}} \oplus \mathbb{C}\dot{K}\right)^*$, set

$$V_\nu^{\mathfrak{g};\mathfrak{p}} = \left\{v \in L(\Lambda) \,|\, Xv = 0 \ (\forall\, X \in \mathfrak{p}_+),\ Hv = \nu(H)\,v \ \left(\forall\, H \in \dot{\overline{\mathfrak{h}}} \oplus \mathbb{C}\dot{K}\right)\right\}.$$

Due to (4) and (6), $V_\nu^{\mathfrak{g};\mathfrak{p}}$ is stable under the actions of $L_n^{\mathfrak{g};\mathfrak{p}}$ $(n \in \mathbb{Z})$. So, $V_\nu^{\mathfrak{g};\mathfrak{p}}$ becomes a *Vir*-module. We call this module the *coset Virasoro module*. Notice that $L(\Lambda)$ is decomposed as a $Vir \oplus [\mathfrak{p},\mathfrak{p}]$-module into

$$L(\Lambda) = \bigoplus_{\nu \in \dot{\overline{\mathfrak{h}}}^* \oplus \mathbb{C}\delta \bmod \mathbb{C}\delta} \left(V_\nu^{\mathfrak{g};\mathfrak{p}} \otimes \dot{L}(\nu)\right), \tag{7}$$

where $\dot{L}(\nu)$ is the irreducible $[\mathfrak{p},\mathfrak{p}]$-module with highest weight ν and $\dot{\delta}$ is identified with $\dot{K}$ via the non-degenerate bilinear form on $\mathfrak{p}$. From (7), we define a function

$$c_{\nu}^{\Lambda}(q) = Tr_{V_{\nu}^{\mathfrak{g};\mathfrak{p}}} q^{-d} = \sum_{j\in\mathbb{Z}_{\geq 0}} \mathrm{mult}_{\Lambda}\left(\nu - j\dot{\delta};\mathfrak{p}\right) q^{j} \left(q = e^{-\delta}\right), \tag{8}$$

where the multiplicity is defined as in ([6], Section 1.6). The function (8) is called the *string function*. Using the string function (8), the decomposition (7) yields the following formula for the character of $L(\Lambda)$:

$$chL(\Lambda) = \sum_{\nu\in\dot{\mathfrak{h}}^{*}\oplus\mathbb{C}\dot{\delta} \bmod \mathbb{C}\dot{\delta}} c_{\nu}^{\Lambda}(q) ch\dot{L}(\nu). \tag{9}$$

Let us now introduce the following numbers:

- $m_{\Lambda} = \frac{|\Lambda+\rho|^{2}}{2(m+h^{\vee})} - \frac{|\rho|^{2}}{2h^{\vee}}$,
- $\dot{m}_{\nu} = \frac{|\nu+\dot{\rho}|^{2}}{2(\dot{m}+h^{\vee})} - \frac{|\dot{\rho}|^{2}}{2h^{\vee}}$,

where $\dot{\rho}$ is the Weyl vector associated with $\mathfrak{p}$.

Then, we define the branching function as $b_{\nu}^{\Lambda}(\tau) = q^{m_{\Lambda}-\dot{m}_{\nu}} c_{\nu}^{\Lambda}(q)$ for $q = e^{2\pi i\tau}$. By the strange formula and (4), we see that the branching function also can be written as $b_{\nu}^{\Lambda}(\tau) = q^{-\frac{1}{24}c^{\mathfrak{g};\mathfrak{p}}} Tr_{V_{\nu}^{\mathfrak{g};\mathfrak{p}}} q^{\ell_0}$ (see [11] (Chapter 12) for the strange formula).

Recall that the normalized character $ch'L(\Lambda)$ is defined as

$$ch'L(\Lambda) = e^{-m_{\Lambda}\delta} chL(\Lambda).$$

Introducing the coordinate (τ, z, t) for $h = 2\pi i(-\tau d + z + tK) \in \mathfrak{h}$, we obtain that $ch'_{L(\Lambda)}(\tau,z,t) = q^{m_{\Lambda}} ch_{L(\Lambda)}(\tau,z,t)$. So, the Formula (9) can be rewritten as

$$ch'_{L(\Lambda)}(\tau,z,t) = \sum_{\nu\in\dot{\mathfrak{h}}^{*}\oplus\mathbb{C}\dot{\delta} \bmod \mathbb{C}\dot{\delta}} b_{\nu}^{\Lambda}(\tau)\, ch'_{\dot{L}(\nu)}(\tau,z,t).$$

4. Tensor Product Decompositions

In this section, we fix an affine Lie algebra $\mathfrak{g} = \left(\mathbb{C}\left[t,t^{-1}\right]\otimes\bar{\mathfrak{g}}\right) \oplus \mathbb{C}K \oplus \mathbb{C}d$ over a finite dimensional simple Lie algebra $\bar{\mathfrak{g}}$. We also fix a Cartan subalgebra $\bar{\mathfrak{h}}$ of $\bar{\mathfrak{g}}$. For $\lambda, \mu \in \mathfrak{h}^{*}$, let $L(\lambda)$ and $L(\mu)$ be irreducible highest weight modules over $\mathfrak{g}$. We denote by π_{λ} and π_{μ} the representations of $\mathfrak{g}$ on $L(\lambda)$ and $L(\mu)$, respectively. Put $m = \lambda(K)$ and $m' = \mu(K)$. Assume that $m + h^{\vee} \neq 0$, $m' + h^{\vee} \neq 0$ and $m + m' + h^{\vee} \neq 0$. It follows from (2) that the Virasoro algebra *Vir* acts on $L(\lambda)$ and $L(\mu)$. The corresponding Virasoro fields are

$$L^{\lambda}(z) = \frac{1}{2(m+h^{\vee})} \sum_{i=1}^{dim\bar{\mathfrak{g}}} :\pi_{\lambda}(u_i(z))\,\pi_{\lambda}\left(u^{i}(z)\right):$$

and

$$L^{\mu}(z) = \frac{1}{2(m'+h^{\vee})} \sum_{i=1}^{dim\bar{\mathfrak{g}}} :\pi_{\mu}(u_i(z))\,\pi_{\mu}\left(u^{i}(z)\right):.$$

Notice that the Virasoro algebra *Vir* acts on $L(\lambda)\otimes L(\mu)$ via the tensor product action

$$L^{\lambda,\mu}(z) = L^{\lambda}(z)\otimes \mathrm{Id}_{L(\mu)} + \mathrm{Id}_{L(\lambda)} \otimes L^{\mu}(z)$$

with the central charge $\frac{m dim\bar{\mathfrak{g}}}{m+h^{\vee}} + \frac{m' dim\bar{\mathfrak{g}}}{m'+h^{\vee}}$.

On the other hand, we may consider the whole tensor product $L(\lambda) \otimes L(\mu)$ as the highest weight $\mathfrak{g}$-module. Applying (2) to the highest weight $\mathfrak{g}$-module $L(\lambda) \otimes L(\mu)$, we get the associated Virasoro field

$$L^{\lambda\otimes\mu}(z) = \frac{1}{2\left(m+m^{'}+h^{\vee}\right)} \sum_{i=1}^{dim\bar{\mathfrak{g}}} \left(\pi_{\lambda} \otimes \pi_{\mu}\right)\left(u_i(z)\right)\left(\pi_{\lambda} \otimes \pi_{\mu}\right)\left(u^i(z)\right)$$

with the central charge $\frac{\left(m+m^{'}\right)dim\bar{\mathfrak{g}}}{m+m^{'}+h^{\vee}}$.

Using (3), we have

$$\begin{aligned} \left[L_m^{\lambda,\mu}, t^n \otimes X\right] &= -nt^{n+m} \otimes X \ \left(X \in \bar{\mathfrak{g}},\ m \in \mathbb{Z}-\{0\},\ n \in \mathbb{Z}\right), \\ \left[L_m^{\lambda\otimes\mu}, t^n \otimes X\right] &= -nt^{n+m} \otimes X \ \left(X \in \bar{\mathfrak{g}},\ m \in \mathbb{Z}-\{0\},\ n \in \mathbb{Z}\right). \end{aligned} \tag{10}$$

Set $\widetilde{L}(z) = L^{\lambda,\mu}(z) - L^{\lambda\otimes\mu}(z) = \sum_{n\in\mathbb{Z}} \widetilde{L}_n z^{-n-2}$. According to ([15], Proposition 10.3), the field $\widetilde{L}(z)$ yields the coset Virasoro field on $L(\lambda) \otimes L(\mu)$ with central charge $\frac{m dim\bar{\mathfrak{g}}}{m+h^{\vee}} + \frac{m^{'} dim\bar{\mathfrak{g}}}{m^{'}+h^{\vee}} - \frac{\left(m+m^{'}\right)dim\bar{\mathfrak{g}}}{m+m^{'}+h^{\vee}}$.

For $\mu \in \bar{\mathfrak{h}}^* \oplus \mathbb{C}\delta$, we define

$$V_{\mu}^{\lambda,\mu} = \left\{v \in L(\lambda) \otimes L(\mu) \,|\, Xv = 0 \ \left(\forall\, x \in \mathfrak{g}_+\right),\ Hv = \mu(H)\, v \ \left(\forall\, H \in \bar{\mathfrak{h}} \oplus \mathbb{C}K\right)\right\}.$$

It follows from (10) that the space $V_{\mu}^{\lambda,\mu}$ becomes a *Vir*-module via the coset Virasoro field $\widetilde{L}(z)$. Notice that $L(\lambda) \otimes L(\mu)$ is decomposed as a $Vir \oplus [\mathfrak{g}, \mathfrak{g}]$-module into

$$L(\lambda) \otimes L(\mu) = \sum_{\mu \in \bar{\mathfrak{h}}^* \oplus \mathbb{C}\delta \bmod \mathbb{C}\delta} V_{\mu}^{\lambda,\mu} \otimes L(\mu). \tag{11}$$

We obtain from (11) a string function

$$c_{\nu}^{\lambda\otimes\mu}(q) = Tr_{V_{\mu}^{\lambda,\mu}} q^{-d} = \sum_{j \in \mathbb{Z}_{\geq 0}} \mathrm{mult}_{\lambda\otimes\mu}\left(\mu - j\delta; \mathfrak{g}\right) q^j. \tag{12}$$

Using (11) and (12), we get

$$ch(\lambda)\, ch(\mu) = \sum_{\mu \in \bar{\mathfrak{h}}^* \oplus \mathbb{C}\delta \bmod \mathbb{C}\delta} c_{\nu}^{\lambda\otimes\mu}(q) chL(\nu). \tag{13}$$

If we define the normalized branching function by

$$b_{\nu}^{\lambda\otimes\mu}(\tau) = q^{m_{\lambda}+m_{\mu}-m_{\nu}} c_{\nu}^{\lambda\otimes\mu}(q),$$

then the Formula (13) yields

$$ch^{'}_{L(\lambda)}(\tau, z, t)\, ch^{'}_{L(\mu)}(\tau, z, t) = \sum_{\nu} b_{\nu}^{\lambda\otimes\mu}(\tau)\, ch^{'}_{L(\nu)}(\tau, z, t). \tag{14}$$

Let Λ be a dominant integral weight and μ a principal admissible weight of the affine Lie algebra $\mathfrak{g}$. Then, the branching function of the tensor product $L(\Lambda) \otimes L(\mu)$ can be expressed in terms of the string functions of $L(\Lambda)$ as follows.

Theorem 3. *Let $\mathfrak{g}$ be any affine Lie algebra and $m \in \mathbb{Z}_{\geq 0}$. Let $m^{'} = \frac{v}{u}$ with $u \in \mathbb{Z}_{\geq 1}$, $v \in \mathbb{Z}$ and $gcd(u,v) = 1$. Assume that Λ and μ^0 are dominant integral weights of level m and $u\left(m^{'}+h^{\vee}\right) - h^{\vee}$,* *respectively. Write* $\widetilde{c}^{\Lambda}_{\xi}(q)$ *for the modified string function* $q^{m_\Lambda - \frac{|\xi|^2}{2m}} c^{\Lambda}_{\xi,(\mathfrak{g};\mathfrak{g})}(q)$ *for* $\xi \in \overline{\mathfrak{h}}^* \oplus \mathbb{C}\delta$*, where* $c^{\Lambda}_{\xi,(\mathfrak{g};\mathfrak{g})}(q)$ *is the string function defined with respect to the pair* $(\mathfrak{g};\mathfrak{g})$ *(i.e.,* $\mathfrak{p} = \mathfrak{g}$ *in (8)). Then, for a principal admissible weight* $\mu = y\left(\mu^0 - (u-1)\left(m' + h^\vee\right)\Lambda_0 + \rho\right) - \rho \in P^{m'}_{u,y}$*, the following formula holds:*

$$ch'_{L(\Lambda)}(\tau,z,t)\, ch'_{L(\mu)}(\tau,z,t) = \sum_{\nu \in P^{m+m'}_{u,y} \text{ s.t. } \nu \equiv \Lambda+\mu \bmod Q} b^{\Lambda\otimes\mu}_{\nu}(\tau)\, ch'_{L(\nu)}(\tau,z,t),$$

where

$$b^{\Lambda\otimes\mu}_{\nu}(\tau) = \sum_{w\in W} \epsilon(w) q^{\frac{\left(m'+h^\vee\right)\left(m+m'+h^\vee\right)}{2m}\left|\frac{w\left(\nu^0+\rho\right)}{m+m'+h^\vee} - \frac{\mu^0+\rho}{m'+h^\vee}\right|^2} \widetilde{c}^{\Lambda}_{y(w(\nu^0+\rho)-(\mu^0+\rho)-(u-1)m\Lambda_0)}(q).$$

Proof. See ([9], Theorem 3.1). □

In the next section, we simply write $\widetilde{c}^{\Lambda}_{\lambda}$ for $\widetilde{c}^{\Lambda}_{\lambda}(q)$ if no confusion seems likely to arise, and will calculate explicitly the branching functions for some specific cases.

5. Explicit Calculations of Branching Functions

Let Λ_0 and Λ_1 be the fundamental weights of $\widehat{\mathfrak{sl}}_2$, and λ a principal admissible weight of $\widehat{\mathfrak{sl}}_2$. In this section, we explicitly calculate the branching functions arising from the tensor product decompositions of $(L(2\Lambda_0) \oplus L(2\Lambda_1)) \otimes L(\lambda)$ and $L(\rho) \otimes L(\lambda)$.

Let us write $\Pi = \{\alpha\}$ for the simple root system of $\mathfrak{sl}_2$. Then it is easy to check

$$h^\vee = 2,\ \rho = \Lambda_0 + \Lambda_1 = \overline{\rho} + h^\vee \Lambda_0 \text{ and } \Lambda_1 = \Lambda_0 + \frac{1}{2}\alpha \tag{15}$$

for $\widehat{\mathfrak{sl}_2}$. Let $m = \frac{v}{u}$ ($u \in 2\mathbb{Z}_{\geq 1}$, $v \in 2\mathbb{Z}+1$), and choose a principal admissible weight λ of level m satisfying $\lambda = \lambda^0 - (u-1)(m+2)\Lambda_0 \in P^m_{u,1}$ for $\lambda^0 \in P^{u(m+2)-2}_+$ (see Theorems 1 and 2).

Applying Theorem 3 to the tensor product representations $L(2\Lambda_0) \otimes L(\lambda)$ and $L(2\Lambda_1) \otimes L(\lambda)$, we obtain

$$\begin{aligned} &ch'_{L(2\Lambda_0)}(\tau,z,t)\, ch'_{L(\lambda)}(\tau,z,t) \\ &= \sum_{\nu \in P^{m+2}_{u,1} \text{ s.t. } \nu \equiv 2\Lambda_0+\lambda \bmod Q} b^{2\Lambda_0\otimes\lambda}_{\nu}(\tau)\, ch'_{L(\nu)}(\tau,z,t), \end{aligned} \tag{16}$$

where

$$b^{2\Lambda_0\otimes\lambda}_{\nu}(\tau) = \sum_{w\in W} \epsilon(w) q^{\frac{(m+2)(m+4)}{4}\left|\frac{w\left(\nu^0+\rho\right)}{m+4} - \frac{\lambda^0+\rho}{m+2}\right|^2} \widetilde{c}^{2\Lambda_0}_{w(\nu^0+\rho)-(\lambda^0+\rho)-2(u-1)\Lambda_0}$$

and

$$\begin{aligned} &ch'_{L(2\Lambda_1)}(\tau,z,t)\, ch'_{L(\lambda)}(\tau,z,t) \\ &= \sum_{\widetilde{\nu} \in P^{m+2}_{u,1} \text{ s.t. } \widetilde{\nu} \equiv 2\Lambda_1+\lambda \bmod Q} b^{2\Lambda_1\otimes\lambda}_{\widetilde{\nu}}(\tau)\, ch'_{L(\widetilde{\nu})}(\tau,z,t), \end{aligned} \tag{17}$$

where

$$b^{2\Lambda_1\otimes\lambda}_{\widetilde{\nu}}(\tau) = \sum_{w\in W} \epsilon(w) q^{\frac{(m+2)(m+4)}{4}\left|\frac{w\left(\widetilde{\nu}^0+\rho\right)}{m+4} - \frac{\lambda^0+\rho}{m+2}\right|^2} \widetilde{c}^{2\Lambda_1}_{w(\widetilde{\nu}^0+\rho)-(\lambda^0+\rho)-2(u-1)\Lambda_0}.$$

Similarly, if we apply Theorem 3 to the tensor product representation $L(\rho) \otimes L(\lambda)$ then we have

$$\begin{aligned} & ch'_{L(\rho)}(\tau,z,t)\, ch'_{L(\lambda)}(\tau,z,t) \\ = & \sum_{\nu \in P^{m+2}_{u,1} \text{ s.t. } \nu \equiv \rho+\lambda \bmod Q} b^{\rho\otimes\lambda}_{\nu}(\tau)\, ch'_{L(\nu)}(\tau,z,t), \end{aligned} \tag{18}$$

where

$$b^{\rho\otimes\lambda}_{\nu}(\tau) = \sum_{w \in W} \epsilon(w) q^{\frac{(m+2)(m+4)}{4}\left|\frac{w(\nu^0+\rho)}{m+4} - \frac{\lambda^0+\rho}{m+2}\right|^2} \tilde{c}^{\rho}_{w(\nu^0+\rho)-(\lambda^0+\rho)-2(u-1)\Lambda_0}.$$

For $\lambda^0 \in P^{u(m+2)-2}_{+}$ and $\nu^0 \in P^{u(m+4)-2}_{+}$, let us write

$$\begin{aligned} \lambda^0 &= (u(m+2)-2-n)\,\Lambda_0 + n\Lambda_1, \\ \nu^0 &= \left(u(m+4)-2-n'\right)\Lambda_0 + n'\Lambda_1 \end{aligned} \tag{19}$$

for some $n \in \mathbb{Z}$ and $n' \in \mathbb{Z}$. Then, we can rewrite λ and ν in (16) as

$$\lambda = \lambda^0 - (u-1)(m+2)\Lambda_0 = (m-n)\Lambda_0 + n\Lambda_1$$

and

$$\nu = \nu^0 - (u-1)(m+4)\Lambda_0 = \left(m-n'+2\right)\Lambda_0 + n'\Lambda_1.$$

Since $2\Lambda_0 - (\nu-\lambda) \in Q$ and $2\Lambda_1 - 2\Lambda_0 = \alpha$, we should have $n \equiv n' \pmod 2$.

Similarly, for $\tilde{\nu}^0 = \left(u(m+4)-2-n''\right)\Lambda_0 + n''\Lambda_1 \in P^{u(m+4)-2}_{+}$, we obtain $\tilde{\nu} = \left(m-n''+2\right)\Lambda_0 + n''\Lambda_1 \left(n'' \in \mathbb{Z}\right)$. From the condition $2\Lambda_1 - (\tilde{\nu}-\lambda) \in \mathbb{Z}\alpha$, we have the same condition $n \equiv n'' \pmod 2$ as the case of ν. For this reason, we shall identify $\tilde{\nu}$ with ν in the following Theorem 4. The same argument yields that the condition $\nu \equiv \rho+\lambda \bmod Q$ in (18) is equivalent to the condition $n' \equiv n+1 \pmod 2$ in (19).

Theorem 4. *Let $m = \frac{v}{u}$ for $u \in 2\mathbb{Z}_{\geq 1}$ and $v \in 2\mathbb{Z}+1$, and let $p = u(m+4)$ and $p' = u(m+2)$.*

1. *Suppose that*

$$\lambda^0 = (u(m+2)-2-n)\,\Lambda_0 + n\Lambda_1$$

and

$$\nu^0 = \left(u(m+2)-2-n'\right)\Lambda_0 + n'\Lambda_1$$

for some $n \in 4\mathbb{Z}$ and $n' \in \mathbb{Z}$ satisfying $n \equiv n' \pmod 2$. Then, the branching functions in (16) and (17) are explicitly given by

$$\begin{aligned} b^{2\Lambda_0\otimes\lambda}_{\nu}(\tau) &= \sum_{j\in\mathbb{Z}} q^{\frac{1}{8pp'}\left(2pp'j+\left(n'+1\right)p'-(n+1)p\right)^2} A \\ &\quad - \sum_{j\in\mathbb{Z}} q^{\frac{1}{8pp'}\left(2pp'j-\left(n'+1\right)p'-(n+1)p\right)^2} B \end{aligned} \tag{20}$$

and

$$\begin{aligned} b_{\nu}^{2\Lambda_1\otimes\lambda}(\tau) &= \sum_{j\in\mathbb{Z}} q^{\frac{1}{8pp'}\left(2pp'j+\left(n'+1\right)p'-(n+1)p\right)^2} B \\ &\quad - \sum_{j\in\mathbb{Z}} q^{\frac{1}{8pp'}\left(2pp'j-\left(n'+1\right)p'-(n+1)p\right)^2} A, \end{aligned} \tag{21}$$

where $\begin{cases} A = \widetilde{c}_{2\Lambda_0}^{2\Lambda_0}, \ B = \widetilde{c}_{2\Lambda_1}^{2\Lambda_0} & \text{if } n' \equiv 0 \pmod 4 \\ A = \widetilde{c}_{2\Lambda_1}^{2\Lambda_0}, \ B = \widetilde{c}_{2\Lambda_0}^{2\Lambda_0} & \text{if } n' \equiv 2 \pmod 4. \end{cases}$

2. *Assume that*

$$\lambda^0 = (u(m+2) - 2 - n)\,\Lambda_0 + n\Lambda_1$$

and

$$\nu^0 = \left(u(m+2) - 2 - n'\right)\Lambda_0 + n'\Lambda_1$$

for some $n \in 4\mathbb{Z}$ *and* $n' \in 4\mathbb{Z}+1$. *Then, the branching function in (18) is explicitly given by*

$$\begin{aligned} & b_{\nu}^{\rho\otimes\lambda}(\tau) \\ &= \sum_{j\in\mathbb{Z}} \left(q^{\frac{1}{8pp'}\left(2pp'j+\left(n'+1\right)p'-(n+1)p\right)^2} - q^{\frac{1}{8pp'}\left(2pp'j-\left(n'+1\right)p'-(n+1)p\right)^2} \right) \widetilde{c}_{\rho}^{\rho}. \end{aligned} \tag{22}$$

Proof. We first prove (20) and (21).

Recall that the Weyl group W of $\widehat{\mathfrak{sl}_2}$ is given by $\{t_{j\alpha},\ t_{j\alpha}r_\alpha | j \in \mathbb{Z}\}$.

By (15) and (19), we have

$$\nu^0 + \rho = u(m+4)\Lambda_0 + \frac{n'+1}{2}\alpha$$

and

$$\lambda^0 + \rho = u(m+2)\Lambda_0 + \frac{n+1}{2}\alpha.$$

So, we get

$$\begin{aligned} & t_{j\alpha}\left(\nu^0+\rho\right) - \left(\lambda^0+\rho\right) - 2(u-1)\Lambda_0 \\ &= 2\Lambda_0 + \left(u(m+4)j + \frac{n'-n}{2}\right)\alpha - \left(u(m+4)j^2 + \left(n'+1\right)j\right)\delta \end{aligned} \tag{23}$$

and

$$\begin{aligned} & t_{j\alpha}r_\alpha\left(\nu^0+\rho\right) - \left(\lambda^0+\rho\right) - 2(u-1)\Lambda_0 \\ &= 2\Lambda_0 + \left(u(m+4)j - \frac{n'+n+2}{2}\right)\alpha + \left(u(m+4)j^2 - \left(n'+1\right)j\right)\delta. \end{aligned} \tag{24}$$

Notice from ([11], (12.7.9)) that we have

$$\widetilde{c}_{w(\lambda')+2\gamma+a\delta}^{2\Lambda_0} = \widetilde{c}_{\lambda'}^{2\Lambda_0} \tag{25}$$

for $\lambda' \in \mathfrak{h}^*$, $w \in \overline{W}$, $\gamma \in \mathbb{Z}\alpha$ and $a \in \mathbb{C}$. Since $\overline{W} = \{1, r_\alpha\}$, we see from (25) that

$$\widetilde{c}_{\lambda+(2n+1)\alpha+a\delta}^{2\Lambda_0} = \widetilde{c}_{(\lambda+\alpha)+2n\alpha+a\delta}^{2\Lambda_0} = \widetilde{c}_{\lambda+\alpha}^{2\Lambda_0}$$

and

$$\widetilde{c}^{2\Lambda_0}_{r_\alpha(\lambda)+(2n+1)\alpha+a\delta} = \widetilde{c}^{2\Lambda_0}_{r_\alpha(\lambda+\alpha)+(2n+2)\alpha+a\delta} = \widetilde{c}^{2\Lambda_0}_{\lambda+\alpha}.$$

Hence, in any case we obtain

$$\widetilde{c}^{2\Lambda_0}_{w(\lambda)+(2n+1)\alpha+a\delta} = \widetilde{c}^{2\Lambda_0}_{\lambda+\alpha} \tag{26}$$

for $w \in \overline{W}$. Since u is even, we have

$$u(m+4)j + \frac{n' - n}{2} \equiv \frac{n' - n}{2} \pmod{2}$$

and

$$u(m+4)j - \frac{n' + n + 2}{2} \equiv -\frac{n' + n + 2}{2} \pmod{2}.$$

Since $n \in 4\mathbb{Z}$ and $n \equiv n' \pmod{2}$, we obtain $n' \equiv 0 \pmod{4}$ or $n' \equiv 2 \pmod{4}$. If $n' \equiv 0 \pmod{4}$, then $\frac{n'-n}{2} \equiv 0 \pmod{2}$ and $-\frac{n'+n+2}{2} \equiv 1 \pmod{2}$. Thus, by (23), (24), (25) and (26) we get

$$\widetilde{c}^{2\Lambda_0}_{t_{j\alpha}(\nu^0+\rho)-(\lambda^0+\rho)-2(u-1)\Lambda_0} = \widetilde{c}^{2\Lambda_0}_{2\Lambda_0} \tag{27}$$

and

$$\widetilde{c}^{2\Lambda_0}_{t_{j\alpha}r_\alpha(\nu^0+\rho)-(\lambda^0+\rho)-2(u-1)\Lambda_0} = \widetilde{c}^{2\Lambda_0}_{2\Lambda_0+\alpha} = \widetilde{c}^{2\Lambda_0}_{2\Lambda_1}. \tag{28}$$

Similarly, if $n' \equiv 2 \pmod{4}$, then $\frac{n'-n}{2} \equiv 1 \pmod{2}$ and $-\frac{n'+n+2}{2} \equiv 0 \pmod{2}$. So, in this case we have

$$\widetilde{c}^{2\Lambda_0}_{t_{j\alpha}(\lambda^0+\rho)-(\mu^0+\rho)-2(u-1)\Lambda_0} = \widetilde{c}^{2\Lambda_0}_{2\Lambda_0+\alpha} = \widetilde{c}^{2\Lambda_0}_{2\Lambda_1} \tag{29}$$

and

$$\widetilde{c}^{2\Lambda_0}_{t_{j\alpha}r_\alpha(\lambda^0+\rho)-(\mu^0+\rho)-2(u-1)\Lambda_0} = \widetilde{c}^{2\Lambda_0}_{2\Lambda_0}. \tag{30}$$

We now compute the exponent $\frac{(m+2)(m+4)}{4}\left|\frac{w(\nu^0+\rho)}{m+4} - \frac{\lambda^0+\rho}{m+2}\right|^2$ of q in (16) and (17).

Since $p = u(m+4)$ in assumption, we see that

$$t_{j\alpha}\left(\nu^0+\rho\right) = p\Lambda_0 + \left(pj + \frac{n'+1}{2}\right)\alpha - \left(pj^2 + \left(n'+1\right)j\right)\delta. \tag{31}$$

and

$$t_{j\alpha}r_\alpha\left(\nu^0+\rho\right) = p\Lambda_0 + \left(pj - \frac{n'+1}{2}\right)\alpha - \left(pj^2 - \left(n'+1\right)j\right)\delta. \tag{32}$$

It also follows from the assumption $p' = u(m+2)$ that

$$\begin{aligned} &\frac{(m+2)(m+4)}{4}\left|\frac{w\left(\nu^0+\rho\right)}{m+4} - \frac{\lambda^0+\rho}{m+2}\right|^2 \\ &= \frac{u^2(m+2)(m+4)}{4}\left|\frac{w\left(\nu^0+\rho\right)}{u(m+4)} - \frac{\lambda^0+\rho}{u(m+2)}\right|^2 \\ &= \frac{pp'}{4}\left|\frac{w\left(\nu^0+\rho\right)}{p} - \frac{\lambda^0+\rho}{p'}\right|^2. \end{aligned} \tag{33}$$

Notice from (31) and (32) that

$$\begin{aligned}
&\frac{t_{j\alpha}\left(\nu^0+\rho\right)}{p}-\frac{\lambda^0+\rho}{p'}\\
=&\left(\Lambda_0+\left(j+\frac{n'+1}{2p}\right)\alpha\right)-\frac{1}{p'}\left(p'\Lambda_0+\frac{n+1}{2}\alpha\right)\ \mathrm{mod}\ \mathbb{C}\delta\\
=&\left(j+\frac{n'+1}{2p}-\frac{n+1}{2p'}\right)\alpha\ \mathrm{mod}\ \mathbb{C}\delta
\end{aligned}$$

and

$$\begin{aligned}
&\frac{t_{j\alpha}r_\alpha\left(\nu^0+\rho\right)}{p}-\frac{\lambda^0+\rho}{p'}\\
=&\left(\Lambda_0+\left(j-\frac{n'+1}{2p}\right)\alpha\right)-\frac{1}{p'}\left(p'\Lambda_0+\frac{n+1}{2}\alpha\right)\ \mathrm{mod}\ \mathbb{C}\delta\\
=&\left(j-\frac{n'+1}{2p}-\frac{n+1}{2p'}\right)\alpha\ \mathrm{mod}\ \mathbb{C}\delta.
\end{aligned}$$

Thus, we obtain

$$\begin{aligned}
\left|\frac{t_{j\alpha}\left(\nu^0+\rho\right)}{p}-\frac{\lambda^0+\rho}{p'}\right|^2&=\frac{1}{2\left(pp'\right)^2}\left(2pp'j+\left(n'+1\right)p'-(n+1)p\right)^2,\\
\left|\frac{t_{j\alpha}r_\alpha\left(\nu^0+\rho\right)}{p}-\frac{\lambda^0+\rho}{p'}\right|^2&=\frac{1}{2\left(pp'\right)^2}\left(2pp'j-\left(n'+1\right)p'-(n+1)p\right)^2.
\end{aligned}\tag{34}$$

Hence, if $n'\equiv 0\pmod 4$, then we obtain from (27), (28), (33) and (34) that

$$\begin{aligned}
b_\nu^{2\Lambda_0\otimes\lambda}(\tau)&=\sum_{j\in\mathbb{Z}}q^{\frac{1}{8pp'}\left(2pp'j+\left(n'+1\right)p'-(n+1)p\right)^2}\widetilde{c}_{2\Lambda_0}^{2\Lambda_0}\\
&\quad-\sum_{j\in\mathbb{Z}}q^{\frac{1}{8pp'}\left(2pp'j-\left(n'+1\right)p'-(n+1)p\right)^2}\widetilde{c}_{2\Lambda_1}^{2\Lambda_0}.
\end{aligned}$$

If $n'\equiv 2\pmod 4$ then we also obtain from (29), (30), (33) and (34)

$$\begin{aligned}
b_\nu^{2\Lambda_0\otimes\lambda}(\tau)&=\sum_{j\in\mathbb{Z}}q^{\frac{1}{8pp'}\left(2pp'j+\left(n'+1\right)p'-(n+1)p\right)^2}\widetilde{c}_{2\Lambda_1}^{2\Lambda_0}\\
&\quad-\sum_{j\in\mathbb{Z}}q^{\frac{1}{8pp'}\left(2pp'j-\left(n'+1\right)p'-(n+1)p\right)^2}\widetilde{c}_{2\Lambda_0}^{2\Lambda_0}.
\end{aligned}$$

The Formula (20) now follows.

Applying the same argument as above to the case of $b_\nu^{2\Lambda_1\otimes\lambda}(\tau)$, we obtain

$$\begin{cases}
\widetilde{c}^{2\Lambda_1}_{t_{j\alpha}(\nu^0+\rho)-(\lambda^0+\rho)-2(u-1)\Lambda_0}=\widetilde{c}^{2\Lambda_1}_{2\Lambda_0} & \text{if } n'\equiv 0\pmod 4\\
\widetilde{c}^{2\Lambda_1}_{t_{j\alpha}r_\alpha(\nu^0+\rho)-(\lambda^0+\rho)-2(u-1)\Lambda_0}=\widetilde{c}^{2\Lambda_1}_{2\Lambda_1} & \text{if } n'\equiv 0\pmod 4.
\end{cases}\tag{35}$$

and

$$\begin{cases} \tilde{c}^{2\Lambda_1}_{t_{j\alpha}(\nu^0+\rho)-(\lambda^0+\rho)-2(u-1)\Lambda_0} = \tilde{c}^{2\Lambda_1}_{2\Lambda_1} & \text{if } n' \equiv 2 \pmod 4 \\ \tilde{c}^{2\Lambda_1}_{t_{j\alpha}r_\alpha(\nu^0+\rho)-(\lambda^0+\rho)-2(u-1)\Lambda_0} = \tilde{c}^{2\Lambda_1}_{2\Lambda_0} & \text{if } n' \equiv 2 \pmod 4). \end{cases} \tag{36}$$

Notice that we have $\tilde{c}^{M\Lambda_0+N\Lambda_1}_{m\Lambda_0+n\Lambda_1} = \tilde{c}^{N\Lambda_0+M\Lambda_1}_{n\Lambda_0+m\Lambda_1}$ due to the outer automorphism of $\widehat{\mathfrak{sl}}_2$.

Hence, we obtain that

$$\tilde{c}^{2\Lambda_1}_{2\Lambda_0} = \tilde{c}^{2\Lambda_0}_{2\Lambda_1} \text{ and } \tilde{c}^{2\Lambda_1}_{2\Lambda_1} = \tilde{c}^{2\Lambda_0}_{2\Lambda_0} \tag{37}$$

Therefore, if $n' \equiv 0 \pmod 4$ then we get from (35), (36), (33), (34) and (37) that

$$\begin{aligned} b^{2\Lambda_1\otimes\lambda}_{\nu}(\tau) &= \sum_{j\in\mathbb{Z}} q^{\frac{1}{8pp'}\left(2pp'j+\left(n'+1\right)p'-(n+1)p\right)^2} \tilde{c}^{2\Lambda_0}_{2\Lambda_1} \\ &\quad - \sum_{j\in\mathbb{Z}} q^{\frac{1}{8pp'}\left(2pp'j-\left(n'+1\right)p'-(n+1)p\right)^2} \tilde{c}^{2\Lambda_0}_{2\Lambda_0}. \end{aligned}$$

Similarly, if $n' \equiv 2 \pmod 4$ then we obtain that

$$\begin{aligned} b^{2\Lambda_1\otimes\lambda}_{\nu}(\tau) &= \sum_{j\in\mathbb{Z}} q^{\frac{1}{8pp'}\left(2pp'j+\left(n'+1\right)p'-(n+1)p\right)^2} \tilde{c}^{2\Lambda_0}_{2\Lambda_0} \\ &\quad - \sum_{j\in\mathbb{Z}} q^{\frac{1}{8pp'}\left(2pp'j-\left(n'+1\right)p'-(n+1)p\right)^2} \tilde{c}^{2\Lambda_0}_{2\Lambda_1}. \end{aligned}$$

The Formula (21) now follows.

Let us now prove (22).

The proof is exactly the same as those of (20) and (21) except for calculations of the string function $\tilde{c}^{\rho}_{w(\nu^0+\rho)-(\lambda^0+\rho)-2(u-1)\Lambda_0}$. Recall from the assumption that $n \in 4\mathbb{Z}$ and $n' \in 4\mathbb{Z}+1$. Then, by (23)–(25) we obtain

$$\tilde{c}^{\rho}_{t_{j\alpha}(\nu^0+\rho)-(\lambda^0+\rho)-2(u-1)\Lambda_0} = \tilde{c}^{\rho}_{2\Lambda_0+\frac{n'-n}{2}\alpha} = \tilde{c}^{\rho}_{2\Lambda_0+\frac{1}{2}\alpha} = \tilde{c}^{\rho}_{\rho}$$

and

$$\tilde{c}^{\rho}_{t_{j\alpha}r_\alpha(\nu^0+\rho)-(\lambda^0+\rho)-2(u-1)\Lambda_0} = \tilde{c}^{\rho}_{2\Lambda_0-\frac{n'+n+2}{2}\alpha} = \tilde{c}^{\rho}_{2\Lambda_0-2\alpha+\frac{1}{2}\alpha} = \tilde{c}^{\rho}_{\rho}.$$

The result now follows. □

It is immediate from Theorem 4 that the branching function of $(L(2\Lambda_0) \oplus L(2\Lambda_1)) \otimes L(\lambda)$ for $\widehat{\mathfrak{sl}}_2$ is given by

$$\begin{aligned} & b^{2\Lambda_0\otimes\lambda}_{\nu}(\tau) + b^{2\Lambda_1\otimes\lambda}_{\nu}(\tau) \\ = & \sum_{j\in\mathbb{Z}} q^{\frac{1}{8pp'}\left(2pp'j+\left(n'+1\right)p'-(n+1)p\right)^2} \left(\tilde{c}^{2\Lambda_0}_{2\Lambda_0} + \tilde{c}^{2\Lambda_0}_{2\Lambda_1}\right) \\ & - \sum_{j\in\mathbb{Z}} q^{\frac{1}{8pp'}\left(2pp'j-\left(n'+1\right)p'-(n+1)p\right)^2} \left(\tilde{c}^{2\Lambda_0}_{2\Lambda_0} + \tilde{c}^{2\Lambda_0}_{2\Lambda_1}\right). \end{aligned}$$

In the following theorem, we explicitly calculate $\tilde{c}^{2\Lambda_0}_{2\Lambda_0} + \tilde{c}^{2\Lambda_0}_{2\Lambda_1}$ and $\tilde{c}^{\rho}_{\rho}$ in terms of the Dedekind eta function.

Theorem 5. $\tilde{c}^{2\Lambda_0}_{2\Lambda_0} + \tilde{c}^{2\Lambda_0}_{2\Lambda_1} = \frac{\eta(\tau)}{\eta\left(\frac{\tau}{2}\right)\eta(2\tau)}$ *and* $\tilde{c}^{\rho}_{\rho} = \frac{\eta(2\tau)}{\eta(\tau)^2}$*, where* $\eta(\tau) = q^{\frac{1}{24}}\prod_{n=1}^{\infty}(1-q^n)$.

Proof. It follows from the Weyl-Kac character formula that

$$chL\,(2\Lambda_0) = \frac{1}{e^{\rho}R}\sum_{w\in W}\epsilon(w)e^{w(2\Lambda_0+\rho)}, \tag{38}$$

where

$$W = \{t_{j\alpha},\ t_{j\alpha}r_\alpha | j \in \mathbb{Z}\}$$

and

$$R = \Pi_{n=1}^{\infty}\left(1-q^n\right)\left(1-e^{-\alpha}q^{n-1}\right)\left(1-e^{\alpha}q^n\right).$$

Calculating $w\,(2\Lambda_0+\rho)$ for $w \in W$, we obtain from (38)

$$chL\,(2\Lambda_0) = \frac{1}{e^{\rho}R}\left(\sum_{j\in\mathbb{Z}} e^{2\Lambda_0+\rho+4j\alpha-\frac{(4j)^2+4j}{4}\delta} - \sum_{j\in\mathbb{Z}} e^{2\Lambda_0+\rho+(-4j-1)\alpha-\frac{(-4j-1)^2+(-4j-1)}{4}\delta}\right). \tag{39}$$

Similarly, we can evaluate $chL\,(2\Lambda_1)$ as follows:

$$\frac{1}{e^{\rho}R}q^{-\frac{1}{2}}\left(\sum_{j\in\mathbb{Z}} e^{2\Lambda_0+\rho+(4j+1)\alpha-\frac{(4j+1)^2+(4j+1)}{4}\delta} - \sum_{j\in\mathbb{Z}} e^{2\Lambda_0+\rho+(-4j-2)\alpha-\frac{(-4j-2)^2+(-4j-2)}{4}\delta}\right). \tag{40}$$

Using (39), (40) and the Jacobi triple product identity, we have

$$\begin{aligned}
& chL\,(2\Lambda_0) - q^{\frac{1}{2}}chL\,(2\Lambda_1) \\
= & \frac{1}{e^{\rho}R}\sum_{j\in\mathbb{Z}}(-1)^j e^{2\Lambda_0+\rho+j\alpha-\frac{j^2+j}{4}\delta} \\
= & \frac{e^{2\Lambda_0}}{R}\sum_{j\in\mathbb{Z}}(-1)^j e^{j\alpha}q^{\frac{j^2+j}{4}} \\
= & \frac{e^{2\Lambda_0}}{R}\prod_{n=1}^{\infty}\left(1-q^{\frac{n}{2}}\right)\left(1-e^{\alpha}q^{\frac{n}{2}}\right)\left(1-e^{-\alpha}q^{\frac{n-1}{2}}\right) \\
= & e^{2\Lambda_0}\prod_{n=1}^{\infty}\left(1-q^{\frac{2n-1}{2}}\right)\left(1-e^{\alpha}q^{\frac{2n-1}{2}}\right)\left(1-e^{-\alpha}q^{\frac{2n-1}{2}}\right) \\
= & e^{2\Lambda_0}\prod_{n=1}^{\infty}\frac{1-q^{n-\frac{1}{2}}}{1-q^n}\prod_{n=1}^{\infty}\left(1-q^n\right)\left(1-e^{\alpha}q^{\frac{2n-1}{2}}\right)\left(1-e^{-\alpha}q^{\frac{2n-1}{2}}\right) \\
= & e^{2\Lambda_0}\prod_{n=1}^{\infty}\frac{1-q^{n-\frac{1}{2}}}{1-q^n}\sum_{j\in\mathbb{Z}}(-1)^j e^{j\alpha}q^{\frac{j^2}{2}}.
\end{aligned} \tag{41}$$

Recall from ([11], (12.7.1)) that

$$chL\,(2\Lambda_0) = \sum_{\lambda\in\mathbf{max}(2\Lambda_0)} c_{\lambda}^{2\Lambda_0}e^{\lambda}. \tag{42}$$

and

$$chL\,(2\Lambda_1) = \sum_{\lambda\in\mathbf{max}(2\Lambda_1)} c_{\lambda}^{2\Lambda_1}e^{\lambda}. \tag{43}$$

From (42) and (43), the coefficient of $e^{2\Lambda_0}$ in $chL\,(2\Lambda_0) - q^{\frac{1}{2}}chL\,(2\Lambda_1)$ should be equal to

$$c_{2\Lambda_0}^{2\Lambda_0} - q^{\frac{1}{2}}c_{2\Lambda_0}^{2\Lambda_1}. \tag{44}$$

Comparing (44) with the coefficient of $e^{2\Lambda_0}$ in (41), we obtain

$$c_{2\Lambda_0}^{2\Lambda_0} - q^{\frac{1}{2}} c_{2\Lambda_0}^{2\Lambda_1} = \prod_{n=1}^{\infty} \frac{1 - q^{n-\frac{1}{2}}}{1 - q^n}. \tag{45}$$

By substituting $x = q^{\frac{1}{2}}$, we obtain from (45)

$$c_{2\Lambda_0}^{2\Lambda_0} - x c_{2\Lambda_0}^{2\Lambda_1} = \prod_{n=1}^{\infty} \frac{1 - x^{2n-1}}{1 - x^{2n}}.$$

By letting $x \longmapsto -x$, we get

$$c_{2\Lambda_0}^{2\Lambda_0} + x c_{2\Lambda_0}^{2\Lambda_1} = \prod_{n=1}^{\infty} \frac{1 + x^{2n-1}}{1 - x^{2n}},$$

and this implies

$$c_{2\Lambda_0}^{2\Lambda_0} + q^{\frac{1}{2}} c_{2\Lambda_0}^{2\Lambda_1} = \prod_{n=1}^{\infty} \frac{1 + q^{n-\frac{1}{2}}}{1 - q^n}. \tag{46}$$

On the other hand, it is easy to check that $m_{2\Lambda_0} - \frac{|2\Lambda_0|^2}{4} = -\frac{1}{16}$ and $m_{2\Lambda_1} - \frac{|2\Lambda_0|^2}{4} = \frac{7}{16}$, and these yield that $\tilde{c}_{2\Lambda_0}^{2\Lambda_0} = q^{-\frac{1}{16}} c_{2\Lambda_0}^{2\Lambda_0}$ and $\tilde{c}_{2\Lambda_0}^{2\Lambda_1} = q^{\frac{7}{16}} c_{2\Lambda_0}^{2\Lambda_1}$. So, (46) gives rise to

$$q^{\frac{1}{16}} \left(\tilde{c}_{2\Lambda_0}^{2\Lambda_0} + \tilde{c}_{2\Lambda_0}^{2\Lambda_1} \right) = \prod_{n=1}^{\infty} \frac{1 + q^{n-\frac{1}{2}}}{1 - q^n}. \tag{47}$$

Thus,

$$\begin{aligned}
& \frac{\eta(\tau)}{\eta\left(\frac{\tau}{2}\right) \eta(2\tau)} \\
= & \frac{q^{-\frac{1}{16}}}{\prod_{n=1}^{\infty} \left(1 - q^{\frac{n}{2}}\right) \prod_{n=1}^{\infty} (1 + q^n)} \\
= & \frac{q^{-\frac{1}{16}} \prod_{n=1}^{\infty} \left(1 + q^{\frac{n}{2}}\right)}{\prod_{n=1}^{\infty} (1 - q^n) \prod_{n=1}^{\infty} (1 + q^n)} \\
= & \frac{q^{-\frac{1}{16}} \prod_{n=1}^{\infty} \left(1 + q^{n-\frac{1}{2}}\right)}{\prod_{n=1}^{\infty} (1 - q^n)} \\
= & \tilde{c}_{2\Lambda_0}^{2\Lambda_0} + \tilde{c}_{2\Lambda_0}^{2\Lambda_1} \quad (\text{see } (47)).
\end{aligned}$$

Next, we compute $\tilde{c}_{\rho}^{\rho}$.

Replacing all positive roots α by $k\alpha$ $(k \in \mathbb{Z}_{\geq 1})$, we obtain from the denominator identity that

$$e^{k\rho} \prod_{\alpha \in \Delta_+} \left(1 - e^{-k\alpha}\right)^{\mathrm{mult}(\alpha)} = \sum_{w \in W} \epsilon(w) e^{w(k\rho)}.$$

Thus, it follows from the Jacobi triple identity that

$$
\begin{aligned}
& chL(\rho) \\
&= \frac{1}{e^{\rho}R}\sum_{w\in W}\epsilon(w)e^{w(2\rho)} \\
&= \frac{1}{e^{\rho}R}e^{2\rho}\prod_{\alpha\in\Delta_+}\left(1-e^{-2\alpha}\right)^{\mathrm{mult}(\alpha)} \\
&= e^{\rho}\prod_{j=1}^{\infty}\frac{(1-q^{2j})\left(1-e^{-2\alpha}q^{2(j-1)}\right)(1-e^{2\alpha}q^{2j})}{(1-q^{j})(1-e^{-\alpha}q^{(j-1)})(1-e^{\alpha}q^{j})} \quad \left(q=e^{-\delta}\right) \\
&= e^{\rho}\prod_{j=1}^{\infty}\frac{(1+q^j)}{(1-q^j)}\prod_{j=1}^{\infty}\left(1-q^j\right)\left(1+e^{-\alpha}q^{(j-1)}\right)\left(1+e^{\alpha}q^j\right) \\
&= \prod_{j=1}^{\infty}\frac{(1+q^j)}{(1-q^j)}\sum_{j\in\mathbb{Z}}e^{\rho-j\alpha}q^{\frac{j^2-j}{2}} \\
&= \prod_{j=1}^{\infty}\frac{(1+q^j)}{(1-q^j)}\sum_{j\in\mathbb{Z}}e^{\rho-j\alpha-\frac{j^2-j}{2}\delta}.
\end{aligned} \tag{48}
$$

On the other hand, we get from ([11], (12.7.1)) that

$$chL(\rho) = \sum_{\lambda\in\mathbf{max}(\rho)} c_{\lambda}^{\rho}e^{\lambda}. \tag{49}$$

Comparing the coefficients of e^{ρ} in (48) and (49), we have

$$c_{\rho}^{\rho} = \prod_{j=1}^{\infty}\frac{1+q^j}{1-q^j}.$$

Moreover, it is easy to check $m_{\rho} - \frac{|\rho|^2}{4} = 0$ which implies $\tilde{c}_{\rho}^{\rho} = c_{\rho}^{\rho}$.

The result now follows. □

6. Super-Virasoro algebras

In this section, we shall investigate relationships between our results on branching functions and the representation theory of super-Virasoro algebras. As by-products, we generalize the tensor product decomposition formulas ([6], (4.1.2a) and (4.1.2b)) to the case of principal admissible weights.

Let us first review the theta functions associated to an affine Lie algebra $\mathfrak{g} = \mathbb{C}\left[t,t^{-1}\right]\otimes\bar{\mathfrak{g}}\oplus\mathbb{C}K\oplus\mathbb{C}d$ and its Cartan subalgebra $\mathfrak{h}$.

For $\lambda \in P^m$ $(m\in\mathbb{Z}_{\geq 0})$, the theta function θ_{λ} is defined as

$$\theta_{\lambda} = e^{-\frac{|\lambda|^2}{2m}\delta}\sum_{\alpha\in\overline{Q}}e^{t_{\alpha}\lambda},$$

where $\overline{Q}$ is the root lattice of $\bar{\mathfrak{g}}$. Using the coordinate (τ,z,t) for the Cartan subalgebra $\mathfrak{h}$, we get

$$\theta_{\lambda}(\tau,z,t) = e^{2\pi imt}\sum_{\gamma\in\overline{Q}+\frac{\overline{\lambda}}{m}}q^{\frac{m}{2}|\gamma|^2}e^{2\pi im(\gamma|z)},$$

where $\overline{\lambda}$ is the projection of λ onto $\bar{\mathfrak{h}}$.

In particular, if we take $\lambda = md + \frac{1}{2}n\alpha + rK \in P^m$ for $\mathfrak{g} = \widehat{\mathfrak{sl}}_2$ then the corresponding theta function is

$$\theta_\lambda(\tau, z, t) = e^{2\pi imt} \sum_{k\in\mathbb{Z}+\frac{n}{m}} q^{mk^2} e^{2\pi im(k\alpha|z)}. \tag{50}$$

Evaluating (50) at $(\tau, 0, 0)$, we have

$$\theta(\tau, 0, 0) = \sum_{j\in\mathbb{Z}} q^{m\left(j+\frac{n}{2m}\right)^2} \quad \left(q = e^{2\pi i\tau}\right). \tag{51}$$

For convenience, we shall simply write $\theta_{n,m}$ for (51) in the remaining part of this section.

Next, we review the super-Virasoro algebras Vir_ϵ $\left(\epsilon = 0, \frac{1}{2}\right)$. (For $\epsilon = 0$ or $\frac{1}{2}$, Vir_ϵ is called the *Ramond* and *Neveu-Schwarz* superalgebra, respectively.)

The super-Virasoro algebra Vir_ϵ is the complex superalgebra with a basis $\{c, \ell_j, g_m | j \in \mathbb{Z} \text{ and } m \in \epsilon + \mathbb{Z}\}$, and it satisfies commutation relations

1. $[\ell_i, \ell_j] = (i-j)\ell_{i+j} + \frac{1}{12}\left(i^3 - i\right)\delta_{i+j,0}c$,
2. $[c, \ell_j] = 0$,
3. $[g_m, \ell_n] = \left(m - \frac{n}{2}\right) g_{m+n}$,
4. $[g_m, c] = 0$,
5. $\{g_m, g_n\} = 2\ell_{m+n} + \frac{1}{3}\left(m^2 - \frac{1}{4}\right)\delta_{m+n,0}c$,

where $\{\ ,\ \}$ denotes an anti-commutator bracket between two odd elements.

Recall that every minimal series irreducible module of Vir_ϵ corresponds to the pair of numbers $\left(z^{(p,p')}, h_{r,s;\epsilon}^{(p,p')}\right)$. Here, $z^{(p,p')}$ is the central charge equals $z^{(p,p')} = \frac{3}{2}\left(1 - \frac{2\left(p-p'\right)^2}{pp'}\right)$, and $h_{r,s;\epsilon}^{(p,p')}$ is the minimal eigenvalue of ℓ_0 equals $h_{r,s;\epsilon}^{(p,p')} = \frac{\left(pr-p's\right)^2 - \left(p-p'\right)^2}{8pp'} + \frac{1}{16}(1-2\epsilon)$ for $p, p', r, s \in \mathbb{Z}$, $2 \le p' < p$, $p - p' \in 2\mathbb{Z}$, $gcd\left(\frac{p-p'}{2}, p'\right) = 1$, $1 \le r \le p' - 1$, $1 \le s \le p - 1$ and $r - s \in 2\mathbb{Z}$ (we refer to ([16], Theorem 5.2) for the details).

Write $V_\epsilon\left(z^{(p,p')}, h_{r,s;\epsilon}^{(p,p')}\right)$ for the minimal series module over Vir_ϵ corresponding to $\left(z^{(p,p')}, h_{r,s;\epsilon}^{(p,p')}\right)$. According to [17,18], it follows that

$$chV_\epsilon\left(z^{(p,p')}, h_{r,s;\epsilon}^{(p,p')}\right) = q^{\frac{1}{24}z^{(p,p')}} \eta_\epsilon(\tau) \left(\theta_{\frac{pr-p's}{2}, \frac{pp'}{2}} - \theta_{\frac{pr+p's}{2}, \frac{pp'}{2}}\right),$$

where $\eta_\epsilon(\tau) = \begin{cases} \frac{\eta(2\tau)}{\eta(\tau)^2} & \text{if } \epsilon = 0 \\ \frac{\eta(\tau)}{\eta\left(\frac{\tau}{2}\right)\eta(2\tau)} & \text{if } \epsilon = \frac{1}{2}. \end{cases}$

By (51), we see that

$$\begin{aligned} \theta_{\frac{pr-p's}{2}, \frac{pp'}{2}} &= \sum_{j\in\mathbb{Z}} q^{\frac{1}{8pp'}\left(2pp'j + pr - p's\right)^2}, \\ \theta_{\frac{pr+p's}{2}, \frac{pp'}{2}} &= \sum_{j\in\mathbb{Z}} q^{\frac{1}{8pp'}\left(2pp'j + pr + p's\right)^2}. \end{aligned}$$

So, the normalized character of $V_\epsilon\left(z^{\left(p,p'\right)},h_{r,s;\epsilon}^{\left(p,p'\right)}\right)$ is

$$\begin{aligned}&\chi_{r,s;\epsilon}^{\left(p,p'\right)}(\tau)\\ =\ &\eta_\epsilon(\tau)\left(\sum_{j\in\mathbb{Z}}q^{\frac{1}{8pp'}\left(2pp'j+pr-p's\right)^2}-\sum_{j\in\mathbb{Z}}q^{\frac{1}{8pp'}\left(2pp'j+pr+p's\right)^2}\right),\end{aligned}\tag{52}$$

where $\chi_{r,s;\epsilon}^{\left(p,p'\right)}(\tau)=q^{-\frac{1}{24}z^{\left(p,p'\right)}}chV_\epsilon\left(z^{\left(p,p'\right)},h_{r,s;\epsilon}^{\left(p,p'\right)}\right)$.

Let $r=-(n+1)$ and $s=\left(n'+1\right)$ in (52). Then, by Theorem 4, Theorem 5 and (52), we obtain the following result.

Proposition 1. *Let $m=\frac{v}{u}$ $(u\in 2\mathbb{Z}_{\geq 1},\ v\in 2\mathbb{Z}+1)$. Suppose that λ is a principal admissible weight of $\widehat{\mathfrak{sl}_2}$ such that $\lambda=\lambda^0-(u-1)(m+2)\Lambda_0\in P_{u,1}^{m}$ for $\lambda^0\in P_+^{u(m+2)-2}$. Then, the branching function $b_\nu^{2\Lambda_0\otimes\lambda}(\tau)+b_\nu^{2\Lambda_1\otimes\lambda}(\tau)$ (resp. $b_\nu^{\rho}(\tau)$) of $\left(L(2\Lambda_0)\oplus L(2\Lambda_1)\right)\otimes L(\lambda)$ (resp. $L(\rho)\otimes L(\lambda)$) is the same as the normalized character $\chi_{-(n+1),n'+1;\frac{1}{2}}^{\left(p,p'\right)}(\tau)$ (resp. $\chi_{-(n+1),n'+1;0}^{\left(p,p'\right)}(\tau)$) of the Neveu-Schwarz (resp. Ramond) superalgebra.*

It follows from Section 4 that

$$\left(L(2\Lambda_0)\oplus L(2\Lambda_1)\right)\otimes L(\lambda)=\sum_\nu\left(V_\nu^{2\Lambda_0,\lambda}\oplus V_\nu^{2\Lambda_1,\lambda}\right)\otimes L(\nu)\tag{53}$$

and

$$L(\rho)\otimes L(\lambda)=\sum_{\nu'}V_{\nu'}^{\rho,\lambda}\otimes L\left(\nu'\right),\tag{54}$$

where ν and ν' are taken over $P_{u,1}^{m+2}$ such that $\nu\equiv 2\Lambda_0+\lambda \bmod Q$ and $\nu'\equiv\rho+\lambda \bmod Q$, respectively.

According to [17] the coset Virasoro action introduced in Section 4 can be extended to the action of super-Virasoro algebras, and (53) and (54) can be considered as decompositions of $V_\epsilon\oplus[\mathfrak{g},\mathfrak{g}]$-module. Thus, (14) and Proposition 1 imply that $V_\nu^{2\Lambda_0}\oplus V_\nu^{2\Lambda_1}$ (resp. V_ν^{ρ}) should be isomorphic to the minimal series module $V_{\frac{1}{2}}\left(z^{p,p'},h_{-(n+1),n'+1;\frac{1}{2}}^{p,p'}\right)$ (resp. $V_0\left(z^{p,p'},h_{-(n+1),n'+1;0}^{p,p'}\right)$) as $Vir_{\frac{1}{2}}$-modules (resp. Vir_0-modules). Hence, we obtain the following theorem.

Theorem 6. *Let m and λ be the same as Proposition 1. Then, we have*

$$\left(L(2\Lambda_0)\oplus L(2\Lambda_1)\right)\otimes L(\lambda)=\sum_\nu V_{\frac{1}{2}}\left(z^{p,p'},h_{-(n+1),n'+1;\frac{1}{2}}^{p,p'}\right)\otimes L(\nu)$$

and

$$L(\rho)\otimes L(\lambda)=\sum_{\nu'}V_0\left(z^{p,p'},h_{-(n+1),n'+1;0}^{p,p'}\right)\otimes L\left(\nu'\right),$$

where ν and ν' are taken over $P_{u,1}^{m+2}$ such that $\nu\equiv 2\Lambda_0+\lambda$ mod Q and $\nu'\equiv\rho+\lambda$ mod Q, respectively.

Funding: This research was supported by the Daegu University Research Grant, 2016.

Conflicts of Interest: The author declares no conflict of interest.

References

1. Friedan, D.; Qiu, Z.; Shenker, S. Conformal invariance, unitarity and critical exponents in two dimensions. *Phys. Rev. Lett.* **1984**, *52*, 1575–1578. [CrossRef]
2. Schwarz, J.H. Superstring theory. *Physics Rep.* **1982**, *83*, 223–322. [CrossRef]
3. Fulton, W.; Harris, J. *Representation Theory: A First Course*; Springer: Berlin/Heidelberg, Germany, 1991; Volume 129.
4. Frenkel, E.; Kac, V.G.; Wakimoto, M. Characters and fusion rules for Walgebras via quantized Drinfeld-Sokolov reduction. *Comm. Math. Phys.* **1992**, *147*, 295–328. [CrossRef]
5. Kac, V.G.; Wakimoto, M. Modular invariant representations of infinite-dimensional Lie algebras and superalgebras. *Proc. Natl. Acad. Sci. USA* **1988**, *85*, 4956–4960. [CrossRef] [PubMed]
6. Kac, V.G.; Wakimoto, M. Modular and conformal invariance constraints in representation theory of affine algebras. *Adv. Math.* **1988**, *70*, 156–236. [CrossRef]
7. Kac, V.G.; Wakimoto, M. Classification of modular invariant representations of affine algebras. In *Infinite-Dimensional Lie Algebras and Groups*; World Scientific: Singapore, 1989; Volume 7, pp. 138–177.
8. Kac, V.G.; Roan, S.S.; Wakimoto, M. Quantum reduction for affine superalgebras. *Comm. Math. Phys.* **2003**, *241*, 307–342. [CrossRef]
9. Kac, V.G.; Wakimoto, M. Branching functions for winding subalgebras and tensor products. *Acta. Appl. Math.* **1990**, *21*, 3–39. [CrossRef]
10. Kac, V.G.; Peterson, D.H. Infinite-dimensional Lie algebras, theta functions and modular forms. *Adv. Math.* **1984**, *53*, 125–264. [CrossRef]
11. Kac, V.G. *Infinite Dimensional Lie Algebras*, 3rd ed.; Cambridge University Press: Cambridge, UK, 1990.
12. Wakimoto, M. Infinite-Dimensional Lie Algebras; In *Translation of Mathematical Monographs*; American Mathematical Society: Providence, RI, USA, 2001; Volume 195.
13. Wakimoto, M. *Lectures on Infinite-Dimensional Lie Algebras*; World Scientific: Singapore, 2001.
14. Kac, V.G. *Vertex Algebras for Beginners*, 2nd ed.; American Mathematical Society: Providence, RI, USA, 1998.
15. Kac, V.G.; Raina, A.K. *Bombay Lectures on Highest Weight Representations*; World Scientific: Singapore, 1987.
16. Iohara, K.; Koga, Y. Representation theory of Neveu-Schwarz and Ramond algebra I: Verma modules. *Adv. Math.* **2003**, *178*, 1–65. [CrossRef]
17. Goddard, P.; Kent, A.; Olive, D. Unitary representations of the Virasoro and super-Virasoro algebras. *Commun. Math. Phys.* **1986**, *103*, 105–119. [CrossRef]
18. Kac, V.G.; Wakimoto, M. Unitarizable Highest Weight Representations of the Virasoro, Neveu-Schwarz and Ramond Algebras. In Proceedings of the Symposium on Conformal Groups and Structures, Lecture Notes in Physics, Clausthal, Germany, 12–14 August 1985; Barut, A.O., Doebner, H.D., Eds.; Springer: Berlin/Heidelberg, Germany, 1986; Volume 261, pp. 345–372.

axioms

MDPI

Article

Cohomology Theory of Nonassociative Algebras with Metagroup Relations

Sergey V. Ludkowski

Department of Applied Mathematics, MIREA—Russian Technological University, av. Vernadsky 78, 119454 Moscow, Russia; sludkowski@mail.ru

Received: 30 April 2019; Accepted: 24 June 2019; Published: 4 July 2019

Abstract: Nonassociative algebras with metagroup relations and their modules are studied. Their cohomology theory is scrutinized. Extensions and cleftings of these algebras are studied. Broad families of such algebras and their acyclic complexes are described. For this purpose, different types of products of metagroups are investigated. Necessary structural properties of metagroups are studied. Examples are given. It is shown that a class of nonassociative algebras with metagroup relations contains a subclass of generalized Cayley–Dickson algebras.

Keywords: nonassociative algebra; cohomology; extension; metagroup

MSC: 16E40; 16D70; 18G60; 17A60; 03C60; 03C90

1. Introduction

Nonassociative algebras comprise a large area of algebra. Among them, Lie algebras and their modifications are widely used in different branches of mathematics and its applications including PDEs, physics, quantum mechanics, informatics, and biology (see, for example, [1–5] and references therein). There are other classes of nonassociative algebras which are less investigated. For example, octonions and generalized Cayley–Dickson algebras play very important roles in mathematics and quantum field theory [6–11]. Their structures and identities have attracted great attention. They are used not only in algebra and noncommutative geometry, but also in noncommutative analysis and PDEs, particle physics, mathematical physics, in the theory of Lie groups and algebras and their generalizations, mathematical analysis, operator theory, and in applications in natural sciences including physics and quantum field theory (see [2,7–9,12–25] and references therein).

A multiplicative law of their canonical generators is nonassociative and leads to a more general notion of a metagroup instead of a group [26]. The preposition meta is used to emphasize that such an algebraic object has milder properties than a group. Their axiomatic metagroups satisfy Conditions (1)–(3) with the weak relation (9), as shown in Definition 1 in Section 2. They were used in [26] to investigate automorphisms and derivations of nonassociative algebras.

An extensive area of investigation of PDEs intersects with cohomologies and deformed cohomologies [27]. Therefore, it is important to develop this area using octonions, Cayley–Dickson algebras, and more general metagroup algebras.

It appears that generators of Cayley–Dickson algebras form objects, which are nonassociative generalizations of groups. They are called metagroups. This means that metagroup algebras include the Cayley–Dickson algebras . This article is devoted to algebras generated by metagroups. Note that a class of metagroups differs substantially from a class of groups. Indeed, a metagroup may be nonassociative, power non-associative, or nonalternative. Moreover, left or right inverse elements in the metagroup may not exist or it may contain elements for which left and right inverse elements do not coincide (see Definition 1 in Section 2).

On the other hand, algebras are frequently studied using cohomology theory. However, the already developed cohomology theory operates with associative algebras. It has been investigated by Hochschild and other authors [1,28–30], but it is not applicable to nonassociative algebras. In some particular cases of nonassociative algebras, such as Lie algebras, pre-Lie algebras, flexible algebras, and alternative algebras, homology theory was developed for the needs of studies of their structures (see, for example, [2,31–33] and references therein). It is necessary to note that classes of these algebras are quite different from classes of generalized Cayley–Dickson algebras and nonassociative algebras with metagroup relations. This work is devoted to the development of cohomology theory for nonassociative algebras, namely for its subclass of algebras with metagroup relations.

Previously, cohomologies of loop spaces on quaternion and octonion manifolds were studied in [17]. They have specific features in comparison with the case of complex manifolds. This is especially caused by the noncommutativity of the quaternion skew field and the nonassociativity of the octonion algebra.

In this article, nonassociative algebras with metagroup relations are studied. Their modules and acyclic complexes are investigated. Their cohomology theory is scrutinized in Section 2. This requires the development of a specific axiomatic model of such algebras and their modules. Necessary structural properties of metagroups are studied in Lemmas 1 and 2. Acyclic complexes and co-chain complexes are described in Proposition 1 and Theorem 1. A relation of the cohomologies with quotient modules is given by Theorem 2. Extensions and cleftings of these algebras are studied in Theorems 3–5 under the framework of cohomology theory. Broad families of such algebras are described. In Theorem 6, inner derivations of nonassociative algebras are investigated. A semisimplicity of nonassociative algebras is investigated in Theorem 7 and Corollary 1.

Different types of products of metagroups are investigated in Theorems 8 and 9 in Section 3. Examples are given. It is shown that a class of nonassociative algebras with metagroup relations contains a subclass of generalized Cayley–Dickson algebras.

All of the key results of this paper are obtained for the first time. They can be used for further studies of nonassociative algebras cohomologies, the structure of nonassociative algebras, operator theory, and the spectral theory of Cayley–Dickson algebras, PDEs, noncommutative analysis, noncommutative geometry, mathematical physics, and their applications in the sciences.

2. Cohomology Theory of Nonassociative Algebras

To avoid misunderstandings we give the necessary definitions.

Definition 1. *Let G be a set with a single-valued binary operation (multiplication) $G^2 \ni (a,b) \mapsto ab \in G$, where G satisfies the following conditions:*

(1) For each a and b in G, there is a unique $x \in G$ with $ax = b$ and

(2) A unique $y \in G$ exists, satisfying $ya = b$, which is denoted by $x = a \setminus b = Div_l(a,b)$ and $y = b/a = Div_r(a,b)$ correspondingly,

(3) A neutral (i.e., unit) element $e_G = e \in G$ exists: $eg = ge = g$ for each $g \in G$.

The set of all elements $h \in G$ commuting and associating with G is

(4) $Com(G) := \{a \in G : \forall b \in G,\ ab = ba\}$,

(5) $N_l(G) := \{a \in G : \forall b \in G, \forall c \in G,\ (ab)c = a(bc)\}$,

(6) $N_m(G) := \{a \in G : \forall b \in G, \forall c \in G,\ (ba)c = b(ac)\}$,

(7) $N_r(G) := \{a \in G : \forall b \in G, \forall c \in G,\ (bc)a = b(ca)\}$,

(8) $N(G) := N_l(G) \cap N_m(G) \cap N_r(G)$;

$\mathcal{C}(G) := Com(G) \cap N(G)$ is called the center $\mathcal{C}(G)$ of G.

We call G a metagroup if a set G possesses a single-valued binary operation and satisfies Conditions (1)–(3) and

(9) $(ab)c = t_3(a,b,c)a(bc)$
for each a, b, and c in G, where $t_3(a,b,c) \in \Psi$, $\Psi \subset \mathcal{C}(G)$;
where t_3 *shortens a notation* $t_{3,G}$, *where* Ψ *denotes a (proper or improper) subgroup of* $\mathcal{C}(G)$.

Then G will be called a central metagroup if, in addition to (9), *it satisfies the condition*

(10) $ab = t_2(a,b)ba$
for each a and b in G, where $t_2(a,b) \in \Psi$.

Particularly, $Inv_l(a) = Div_l(a,e)$ is a left inversion, and $Inv_r(a) = Div_r(a,e)$ is a right inversion.

In view of the nonassociativity of G, in general, a product of several elements of G is usually specified by opening "(" and closing ")" parentheses. We denote the product of elements $a_1,...,a_n$ in G by $\{a_1,...,a_n\}_{q(n)}$, where a vector $q(n)$ indicates an order of pairwise multiplications of elements in the row $a_1,...,a_n$ in braces in the following manner. The enumerate positions are as follows: before a_1 by 1, between a_1 and a_2 by 2,..., by n between a_{n-1} and a_n, and by $n+1$ after a_n. Then, we put $q_j(n) = (k,m)$ if there are k opening "(" and m closing ")" parentheses in the ordered product at the j-th position of the type)...)(...(, where k and m are nonnegative integers, $q(n) = (q_1(n),....,q_{n+1}(n))$ with $q_1(n) = (k,0)$ and $q_{n+1}(n) = (0,m)$.

Traditionally, S_n denotes the symmetric group of the set $\{1,2,...,n\}$. Henceforth, maps and functions on metagroups are assumed to be single-valued unless otherwise specified .

Let $\psi : G \to G$ be a bijective surjective map satisfying the following condition: $\psi(ab) = \psi(a)\psi(b)$ for each a and b in G. Then, ψ is called an automorphism of the metagroup G.

Lemma 1. (*i*). *Let G be a central metagroup. Then, for every* $a_1,...,a_n$ *in G,* $v \in S_n$ *and vectors* $q(n)$ *and* $u(n)$ *indicating an order of pairwise multiplications and* $n \in \mathbf{N}$, *there exists an element* $t_n = t_n(a_1,...,a_n;q(n),u(n)|v) \in \Psi$ *such that*

(1) $\{a_1,...,a_n\}_{q(n)} = t_n\{a_{v(1)},...,a_{v(n)}\}_{u(n)}$.

(*ii*). *If G is a metagroup and if v is the neutral element* $v = id$ *in* S_n, *then property* (1) *is satisfied.*

Proof. From Conditions (1)–(8) in Definition 1, it follows that $\mathcal{C}(G)$ itself is a commutative group.

(*i*). For $n=1$, evidently $t_1 = 1$, since $a = 1a$ for each $a \in G$. For $n = 2$, Formula (1) is a direct consequence of condition (10) in Definition 1. Consider $n = 3$. When u is the identity element of S_3, the statement follows from condition (9) in Definition 1. For any transposition u of two elements of the set $\{1,2,3\}$, the statement follows from (9) and (10) in Definition 1. Elements of S_3 can be obtained by multiplication of pairwise transpositions. Therefore, from the condition $\Psi \subset \mathcal{C}(G)$, it follows that formula (1) is valid.

Now, let $n \geq 4$ and suppose that this lemma is proved for any products consisting of less than n elements. In view of Properties (1) and (2) in Definition 1, it is sufficient to verify Formula (1) of this lemma for $\{a_1,...,a_n\}_{q(n)} = (...((a_1a_2)a_3)...)a_n =: \{a_1,...,a_n\}_{l(n)}$ since $\Psi \subset \mathcal{C}(G)$. In this particular case, $\{a_{v(1)},...,a_{v(n)}\}_{u(n)} = \{a_{v(1)},...,a_{v(n-1)}\}_{u(n-1)}a_n$. Formula (1) follows from the induction hypothesis, since $(...((a_1a_2)a_3)...)a_{n-1} = t_{n-1}\{a_{v(1)},...,a_{v(n-1)}\}_{u(n-1)}$ and hence $((...((a_1a_2)a_3)...)a_{n-1})a_n = t_{n-1}(\{a_{v(1)},...,a_{v(n-1)}\}_{u(n-1)}a_n)$ and putting $t_n = t_{n-1}$, where $t_{n-1} = t_{n-1}(a_1,...,a_{n-1};q(n-1),u(n-1)|w)$ with $w = v|_{\{1,...,n-1\}}$, $v(n) = n$.

In the general case, $\{a_{v(1)},...,a_{v(n)}\}_{u(n)} = \{b_1,...,b_j,...,b_k\}_{p(k)}$, where j is such that either $b_j = c_ja_n$ with $c_j = \{a_{v(j)},...,a_{v(j+m-1)}\}_{r(m)}$ and with $v(j+m) = n$ or $b_j = a_nc_j$ with $c_j = \{a_{v(j+1)},...,a_{v(j+m)}\}_{r(m)}$ and with $v(j) = n$. Also, $b_1 = a_{v(1)},...,b_{j-1} = a_{v(j-1)}, b_{j+1} = a_{v(j+1)},..., b_k = a_{v(n)}$ with suitable vectors $p(k)$ and $r(m)$. If $m > 1$, then $k < n$ and using the induction hypothesis for $\{b_1,...,b_j,...,b_k\}_{p(k)}$ and b_j, we get that elements s and t in Ψ exist so that $\{b_1,...,b_j,...,b_k\}_{p(k)} = s\{b_1,...,b_{j-1},b_{j+1},,...,b_k\}_{p(k-1)}b_j$ $= st(\{b_1,...,b_{j-1},b_{j+1},,...,b_k\}_{p(k-1)}c_j)a_n$, where $p(k-1)$ is a corresponding vector prescribing an order of multiplications.

Again, applying the induction hypothesis to the product of $n-1$ elements $\{b_1,...,b_{j-1},b_{j+1},,...,b_k\}_{p(k-1)}c_j$, we deduce that there exists $w \in \Psi$, such that

$\{a_{v(1)},...,a_{v(n)}\}_{u(n)} = stw((...((a_1a_2)a_3)...)a_{n-1})a_n$.

Therefore, a case remains when $m = 1$. Let the first multiplication in $\{a_{v(1)},...,a_{v(n)}\}_{u(n)}$ containing a_n be $(a_{v(k)}a_{v(k+1)}) =: b_k$. We put $b_j = a_{v(j)}$ for each $1 \leq j \leq k-1$. We also put $b_{j-1} = a_{v(j)}$ for each $k+1 < j \leq n$, where either $a_n = a_{v(k)}$ or $a_n = a_{v(k+1)}$. Therefore, using previous identities, we rewrite the considered product as $\{a_{v(1)},...,a_{v(n)}\}_{u(n)} = \{b_{y(1)},...,b_{y(n-1)}\}_{w(n-1)}$ with an element $y \in S_{n-1}$ of the symmetric group and a vector $w(n-1)$, indicating an order of pairwise products (see Definition 1). From the induction hypothesis, we deduce that there exists $t_{n-1} \in \Psi$, so that $t_{n-1}\{b_{y(1)},...,b_{y(n-1)}\}_{w(n-1)} = pb_k$ with $p = \{b_1,...,b_{k-1},b_{k+1},...,b_{n-1}\}_{w(n-1)}$, because G is the central metagroup. Applying the induction hypothesis for $n = 3$, we infer that $t_3 \in \Psi$ exists, such that $t_{n-1}t_3\{b_{y(1)},...,b_{y(n-1)}\}_{w(n-1)} = (pa)a_n$, where either $a = a_{v(k+1)}$ or $a = a_{v(k)}$, correspondingly. From the induction hypothesis for $n-1$, it follows that $\tilde{t}_{n-1} \in \Psi$ exists, so that $\tilde{t}_{n-1}pa = (...((a_1a_2)a_3)...)a_{n-1}$, and hence, $\{a_1,...,a_n\}_{l(n)} = t_n\{a_{v(1)},...,a_{v(n)}\}_{u(n)}$, where $t_n = \tilde{t}_{n-1}t_{n-1}t_3$.

(ii). Now, let G be a metagroup and $v = id$ be the neutral element of the symmetric group S_n, where $id(k) = k$ for each $k \in \mathbf{N}$. Then, using condition (10) of Definition 1 is unnecessary, because transpositions are already not utilized. For $n = 1$ and $n = 2$, we get $t_1 = 1$ and $t_2 = 1$, since $a = 1a$ and $ab = 1ab$ for each a and b in G. For $n = 3$, Formula (1) of this lemma follows on from condition (9) in Definition 1. Then, the proof in case (ii) by induction is a simplification of that of case (i). □

Lemma 2. *If G is a metagroup, then for each a and $b \in G$, the following identities are fulfilled:*

(1) $b \setminus e = (e/b)t_3(e/b,b,b \setminus e)$;

(2) $(a \setminus e)b = (a \setminus b)t_3(e/a,a,a \setminus e)/t_3(e/a,a,a \setminus b)$;

(3) $b(e/a) = (b/a)t_3(b/a,a,a \setminus e)/t_3(e/a,a,a \setminus e)$.

Proof. Conditions (1)–(3) in Definition 1 imply that

(4) $b(b \setminus a) = a,\ b \setminus (ba) = a$;

(5) $(a/b)b = a,\ (ab)/b = a$

for each a and b in G. Using Condition (9) in Definition 1 and Identities (4) and (5), we deduce that $e/b = (e/b)(b(b \setminus e)) = (b \setminus e)/t_3(e/b,b,b \setminus e)$ which leads to (1).

Let $c = a \setminus b$. Then, from Identities (1) and (4), it follows that $(a \setminus e)b = (e/a)t_3(e/a,a,a \setminus e)(ac)$ $= ((e/a)a)(a \setminus b)t_3(e/a,a,a \setminus e)/t_3(e/a,a,a \setminus b)$ which provides (2).

Now, let $d = b/a$. Then, Identities (1) and (5) imply that $b(e/a) = (da)(a \setminus e)/t_3(e/a,a,a \setminus e) =$ $(b/a)t_3(b/a,a,a \setminus e)/t_3(e/a,a,a \setminus e)$ which demonstrates (3). □

Definition 2. *Let A be an algebra over an associative unital ring $\mathcal{T}$, such that A has a natural structure of a $(\mathcal{T},\mathcal{T})$-bimodule with a multiplication map $A \times A \to A$, which is right and left distributive, $a(b+c) = ab + ac$, $(b+c)a = ba + ca$, and also satisfies the identities $r(ab) = (ra)b$, $(ar)b = a(rb)$, $(ab)r = a(br)$, $s(ra) = (sr)a$, and $(ar)s = a(rs)$ for any a, b, and c in A, r and s in $\mathcal{T}$. Let G be a metagroup and $\mathcal{T}$ be an associative unital ring.*

Henceforth, the ring $\mathcal{T}$ is assumed to be commutative, unless otherwise specified.

Then, by $\mathcal{T}[G]$, a metagroup algebra is denoted over $\mathcal{T}$ for all formal sums $s_1a_1 + ... + s_na_n$ satisfying Conditions $(1-3)$ below, where n is a positive integer, s_1,...,s_n are in $\mathcal{T}$, and a_1,...,a_n belong to G:

(1) *$sa = as$ for each s in $\mathcal{T}$ and a in G,*

(2) *$s(ra) = (sr)a$ for each s and r in $\mathcal{T}$, and $a \in G$,*

(3) *$r(ab) = (ra)b$, $(ar)b = a(rb)$, $(ab)r = a(br)$ for each a and b in G, $r \in \mathcal{T}$.*

Note 1. *Let M be an additive commutative group such that M is a two-sided G-module, where G is a metagroup. We remind the reader that this means that automorphisms $p(g)$ and $s(g)$ of M correspond to each $g \in G$. For short, we use $gx = p(g)x$ and $xg = xs(g)$ for each $g \in G$.*

Note that, usually, M has a natural structure of a two-sided $\mathbf{Z}$-module, because M is the additive commutative group, where $\mathbf{Z}$ denotes the ring of all integers. Therefore, M is a two-sided G-module if and only if it is a two-sided $\mathbf{Z}[G]$-module according to the formulas $(\sum_{g\in G} n(g)g)x = \sum_{g\in G} n(g)(gx)$ and $x(\sum_{g\in G} n(g)g) = \sum_{g\in G}(xg)n(g)$, where $n(g) \in \mathbf{Z}$ for each $g \in G$.

One can consider the additive group of integers $\mathbf{Z}$ as the trivial two-sided G-module putting $gn = ng = n$ for each $g \in G$ and $n \in \mathbf{Z}$, where G is a metagroup.

Example 1. *I. Recall the following: Let A be a unital algebra over a commutative associative unital ring F supplied with a scalar involution $a \mapsto \bar{a}$ so that its norm N and trace T maps have values in F and fulfill conditions:*

(1) $a\bar{a} = N(a)1$ *with* $N(a) \in F$,
(2) $a + \bar{a} = T(a)1$ *with* $T(a) \in F$,
(3) $T(ab) = T(ba)$

for each a and b in A.

If a scalar $f \in F$ satisfies the condition $\forall a \in A\ fa = 0 \Rightarrow a = 0$, then such element f is called cancelable. For a cancelable scalar f, the Cayley–Dickson doubling procedure provides new algebra $C(A, f)$ over F such that

(4) $C(A, f) = A \oplus Al$,
(5) $(a + bl)(c + dl) = (ac - f\bar{d}b) + (da + b\bar{c})l$ *and*
(6) $\overline{(a + bl)} = \bar{a} - bl$

for each a and b in A. Then, l is called a doubling generator. From the definitions of T and N, it follows that $\forall a \in A, \forall b \in A\ T(a) = T(a + bl)$ and $N(a + bl) = N(a) + fN(b)$. The algebra A is embedded into $C(A, f)$ as $A \ni a \mapsto (a, 0)$, where $(a, b) = a + bl$. This is put by induction $A_n(f_{(n)}) = C(A_{n-1}, f_n)$, where $A_0 = A$, $f_1 = f$, $n = 1, 2, ...$, $f_{(n)} = (f_1, ..., f_n)$. Then, $A_n(f_{(n)})$ are generalized Cayley–Dickson algebras when F is not a field or Cayley–Dickson algebras when F is a field.

It is natural to put $A_\infty(f) := \bigcup_{n=1}^{\infty} A_n(f_{(n)})$, where $f = (f_n : n \in \mathbf{N})$. If $char(F) \neq 2$, let $Im(z) = z - T(z)/2$ be the imaginary part of a Cayley–Dickson number z, and hence $N(a) := N_2(a, \bar{a})/2$, where $N_2(a, b) := T(a\bar{b})$.

If the doubling procedure starts from $A = F1 =: A_0$, then $A_1 = C(A, f_1)$ is a $*$-extension of F. If A_1 has a basis $\{1, u\}$ over F with the multiplication table $u^2 = u + w$, where $w \in F$ and $4w + 1 \neq 0$ with the involution $\bar{1} = 1$, $\bar{u} = 1 - u$, then A_2 is the generalized quaternion algebra, and A_3 is the generalized octonion (Cayley–Dickson) algebra.

When $F = \mathbf{R}$ and $f_n = 1$, each n by $\mathcal{A}_r$ will denote the real Cayley–Dickson algebra with generators $i_0, ..., i_{2^r-1}$, such that $i_0 = 1$, $i_j^2 = -1$ for each $j \geq 1$, $i_j i_k = -i_k i_j$ for each $j \neq k \geq 1$. Note that the Cayley–Dickson algebra $\mathcal{A}_r$ for each $r \geq 3$ is nonassociative, for example, $(i_1 i_2) i_4 = -i_1(i_2 i_4)$, etc. Moreover, for each $r \geq 4$, the Cayley–Dickson algebra $\mathcal{A}_r$ is nonalternative (see [7–9]). Frequently, $\bar{a}$ is also denoted by a^* or $\tilde{a}$. Then, $G_r = \{i_j, -i_j : j = 0, 1, ..., 2^r - 1\}$ is a finite metagroup for each $3 \leq r < \infty$.

Let A_n be a Cayley–Dickson algebra over a commutative associative unital ring $\mathcal{R}$ that is characteristically different from two, such that $A_0 = \mathcal{R}$, $n \geq 2$. Take its basic generators $i_0, i_1, ..., i_{2^n-1}$, where $i_0 = 1$. Choose $\mathbf{\Psi}$ as a multiplicative subgroup contained in the ring $\mathcal{R}$, such that $f_j \in \mathbf{\Psi}$ for each $j = 0, ..., n$. Put $G_n = \{i_0, i_1, ..., i_{2^n-1}\} \times \mathbf{\Psi}$. Then, G_n is a central metagroup.

II. More generally, let H be a group such that $\mathbf{\Psi} \subset H$ with relations $hi_k = i_k h$ and $(hg)i_k = h(gi_k)$ for each $k = 0, 1, ..., 2^n - 1$ and each h and g in H. Then, $G_n = \{i_0, i_1, ..., i_{2^n-1}\} \times H$ is also a metagroup. If the group H is noncommutative, then the latter metagroup can be noncentral (see Condition (10) in

Definition 1). Using the notation of Example 1. I, we get that the Cayley–Dickson algebra $\mathcal{A}_\infty$ over the real field $\mathbf{R}$ with $f_n = 1$ for each n provides an example of a metagroup $G_\infty = \{i_j, \ -i_j : 0 \le j \in \mathbf{Z}\}$, where $\mathbf{Z}$ denotes the ring of integers.

III. Generally metagroups need not be central. From given metagroups, new metagroups can be constructed using their direct or semidirect products. Certainly, each group is a metagroup also. Therefore, there are abundant families of noncentral metagroups and also of central metagroups different from groups.

In another way, smashed products of groups and of metagroups can be considered by providing other examples of metagroups (for more detail, see Section 3).

Definition 3. *Let $\mathcal{R}$ be a ring, which may be nonassociative relative to the multiplication. If the mapping $\mathcal{R} \times M \to M$, $\mathcal{R} \times M \ni (a, m) \mapsto am \in M$ exists, such that $a(m+k) = am + ak$ and $(a+b)m = am + bm$ for each a and b in $\mathcal{R}$, m and k in M, then M will be called a generalized left $\mathcal{R}$-module or, for short, a left $\mathcal{R}$-module or left module over $\mathcal{R}$.*

If $\mathcal{R}$ is a unital ring and $1m = m$ for each $m \in M$, then M is called a left unital module over $\mathcal{R}$, where 1 denotes the unit element in the ring $\mathcal{R}$. Symmetrically, a right $\mathcal{R}$-module is defined.

If M is a left and right $\mathcal{R}$-module, then it is called a two-sided $\mathcal{R}$-module or a $(\mathcal{R}, \mathcal{R})$-bimodule. If M is a left $\mathcal{R}$-module and a right $\mathcal{S}$-module, then it is called a $(\mathcal{R}, \mathcal{S})$-bimodule.

A two-sided module M over $\mathcal{R}$ is called cyclic if an element $y \in M$ exists such that $M = \mathcal{R}(y\mathcal{R})$ and $M = (\mathcal{R}y)\mathcal{R}$, where $\mathcal{R}(y\mathcal{R}) = \{s(yp) : \ s, p \in \mathcal{R}\}$ and $(\mathcal{R}y)\mathcal{R} = \{(sy)p : \ s, p \in \mathcal{R}\}$.

Let G be a metagroup. Take a metagroup algebra $A = \mathcal{T}[G]$ and a two-sided A-module M, where $\mathcal{T}$ is an associative unital ring (see Definition 2). Let M_g be a two-sided $\mathcal{T}$-module for each $g \in G$, where G is the metagroup. Let M have the decomposition $M = \sum_{g \in G} M_g$ as a two-sided $\mathcal{T}$-module. Let M also satisfy the following conditions:

(1) $hM_g = M_{hg}$ and $M_g h = M_{gh}$,
(2) $(bh)x_g = b(hx_g)$ and $x_g(bh) = (x_g h)b$ and $bx_g = x_g b$,
(3) $(hs)x_g = t_3(h,s,g)h(sx_g)$ and $(hx_g)s = t_3(h,g,s)h(x_g s)$ and $(x_g h)s = t_3(g,h,s)x_g(hs)$

for every h, g, s in G and $b \in \mathcal{T}$ and $x_g \in M_g$. Then, a two-sided A-module M satisfying conditions (1)–(3) is called smashly G-graded. For short, it is also called "G-graded" instead of "smashly G-graded". In particular, if the module M is G-graded and splits into a direct sum $M = \bigoplus_{g \in G} M_g$ of two-sided $\mathcal{T}$-submodules M_g, then we say that that M is directly G-graded. For a nontrivial (nonzero) G-graded module X with the nontrivial metagroup G, it is supposed that $g \in G$ exists such that $X_g \ne X_e$ if something else is not outlined.

Similarly, G-graded left and right A-modules are defined. Henceforward, speaking about A-modules (left, right, or two-sided), it is supposed that they are G-graded and, for short, "an A-module" is written instead of "a G-graded A-module", unless otherwise specified.

If P and N are left A-modules and a homomorphism $\gamma : P \to N$ is such that $\gamma(ax) = a\gamma(x)$ for each $a \in A$ and $x \in P$, then γ is called a left A-homomorphism. Analogously, right A-homomorphisms are defined for right A-modules. For two-sided A modules, a left and right A-homomorphism is called an A-homomorphism.

For left $\mathcal{T}$-modules M and N by $Hom_{\mathcal{T}}(M, N)$, a family of all left $\mathcal{T}$-homomorphisms is defined from M into N. A similar notation is used for a family of all $\mathcal{T}$-homomorphisms (or right $\mathcal{T}$-homomorphisms) of two-sided $\mathcal{T}$-modules (or right $\mathcal{T}$-modules correspondingly). If an algebra A is specified, a homomorphism may be written for short, instead of an A-homomorphism.

Example 2. *Let $\mathcal{T}$ be a commutative associative unital ring. Also, let G be a metagroup and $A = \mathcal{T}[G]$ be a metagroup algebra, where A is considered to be a $\mathcal{T}$-algebra. Put $K_{-1} = A$, $K_0 = A \otimes_{\mathcal{T}} A$ and use induction $K_{n+1} = K_n \otimes_{\mathcal{T}} A$ for each natural number n. Each K_n is supplied with a two-sided A-module structure:*

(1) $\forall p \in \mathcal{T}[C(G)],\ p \cdot (x_0, ..., x_{n+1}) = ((px_0), ..., x_{n+1})$ *and*

$(x_0, ..., (x_{n+1}p)) = (x_0, ..., x_{n+1}) \cdot p$ *and*

$\forall j \in \{1, ..., n\},\ p \cdot (x_0, ..., x_{n+1}) = (x_0, ..., (px_j), ..., x_{n+1})$ *and*

$(x_0, ..., (x_j p), ..., x_{n+1}) = (x_0, ..., x_{n+1}) \cdot p,$

where $0 \cdot (x_1, ..., x_n) = 0;$

(2) $(xy) \cdot (x_0, ..., x_{n+1}) = t_3 \cdot (x \cdot (y \cdot (x_0, ..., x_{n+1})))$

with $t_3 = t_3(x, y, b)$, *(see also Formula (9) in Definition 1 above);*

(3) $t_3 \cdot ((x_0, ..., x_{n+1}) \cdot (xy)) = ((x_0, ..., x_{n+1}) \cdot x) \cdot y$ *with* $t_3 = t_3(b, x, y);$

(4) $(x \cdot (x_0, ..., x_{n+1})) \cdot y = t_3 \cdot (x \cdot ((x_0, ..., x_{n+1}) \cdot y))$ *with* $t_3 = t_3(x, b, y);$

(5) $x \cdot (x_0, ..., x_{n+1}) = t_{n+3}(x, x_0, ..., x_{n+1}; v_0(n+3); l(n+3)) \cdot ((xx_0), x_1, ..., x_{n+1})$
where $\{x, x_0, ..., x_{n+1}\}_{v_0(n+3)} = x\{x_0, ..., x_{n+1}\}_{l(n+2)},$

$\{x_0, ..., x_{n+1}\}_{l(n+2)} = \{x_0, ..., x_n\}_{l(n+1)} x_{n+1},$

$\{x_0\}_{l(1)} = x_0,\ \{x_0 x_1\}_{l(2)} = x_0 x_1;$
where $b = \{x_0, ..., x_{n+1}\}_{l(n+2)},$

$t_n(x_1, ..., x_n; u(n), w(n)) := t_n(x_1, ..., x_n; u(n), w(n)|id)$
using the shortened notation;

(6) $(x_0, ..., x_{n+1}) \cdot x = t_{n+3}(x_0, ..., x_{n+1}, x; l(n+3), v_{n+2}(n+3)) \cdot (x_0, ..., x_n, (x_{n+1}x))$
for every $x, y, x_0, ..., x_{n+1}$ *in* G, *where* $(x_0, ..., x_{n+1})$ *denotes a basic element of* K_n *over* $\mathcal{T}$, *corresponding to the left ordered tensor product*

$(...((x_0 \otimes x_1) \otimes x_2) ... \otimes x_n) \otimes x_{n+1},$

$\{x_0, ..., x_{n+1}, x\}_{v_{n+2}(n+3)} = \{x_0, ..., x_n, x_{n+1}x\}_{l(n+2)}.$

Proposition 1. *For each metagroup algebra* $A = \mathcal{T}[G]$ *(see Definition 2), an acyclic left* A*-complex* $\mathcal{K}$ *exists.*

Proof. Take two-sided A-modules K_n, as in example 2. We construct a boundary $\mathcal{T}$-linear operator $\partial_n : K_n \to K_{n-1}$ on K_n. For basic elements, it is given by the following formulas:

(1) $\partial_n((x \cdot (x_0, x_1, ..., x_n, x_{n+1})) \cdot y) =$

$\sum_{j=0}^{n} (-1)^j \cdot t_{n+4}(x, x_0, ..., x_{n+1}, y; l(n+4), u_{j+1}(n+4))$

$\cdot ((x \cdot (< x_0, x_1, ..., x_{n+1} >_{j+1,n+2})) \cdot y)$, where

(2) $< x_0, ..., x_{n+1} >_{1,n+2} := ((x_0 x_1), x_2, ..., x_{n+1}),$

(3) $< x_0, ..., x_{n+1} >_{2,n+2} := (x_0, (x_1 x_2), x_3, ..., x_{n+1}), ...,$

(4) $< x_0, ..., x_{n+1} >_{n+1,n+2} := (x_0, ..., x_{n-1}, (x_n x_{n+1})),$

(5) $\partial_0(x \cdot (x_0, x_1)) \cdot y = (x \cdot (x_0 x_1)) \cdot y,$

(6) $\{x_0, x_1, ..., x_{n+1}\}_{l(n+2)} := (...((x_0 x_1) x_2) ...) x_{n+1};$

(7) $\{x, x_0, ..., x_{n+1}, y\}_{u_1(n+4)} := (x\{(x_0 x_1), x_2, ..., x_{n+1}\}_{l(n+1)}) y, ...,$

(8) $\{x, x_0, ..., x_{n+1}, y\}_{u_{n+1}(n+4)} := (x\{x_0, x_1, ..., (x_n x_{n+1})\}_{l(n+1)}) y$

for each $x, x_0, ..., x_{n+1}, y$ in G. On the other hand, from formulas (1) and (2) in Definition 1, it follows that $t_{n+4}(x, x_0, ..., x_{n+1}, y; l(n+4), u_{j+1}(n+4)) = t_{n+2}(x_0, ..., x_{n+1}; l(n+2), v_{j+1}(n+2))$ for each $j = 0, ..., n$, where

(9) $\{x_0, ..., x_{n+1}\}_{v_1(n+2)} := \{(x_0 x_1), x_2, ..., x_{n+1}\}_{l(n+1)}, ...,$

(10) $\{x_0, ..., x_{n+1}\}_{v_{n+1}(n+2)} := \{x_0, x_1, ..., (x_n x_{n+1})\}_{l(n+1)}$

for every $x_0, ..., x_{n+1}$ in G. Therefore, ∂_n is a left and right A-homomorphism of (A, A)-modules. In particular, $\partial_1((x \cdot (x_0, x_1, x_2)) \cdot y) = (x \cdot ((x_0 x_1), x_2)) \cdot y - t_3(x_0, x_1, x_2) \cdot (x \cdot (x_0, (x_1 x_2))) \cdot y,$

$\partial_2((x \cdot (x_0, x_1, x_2, x_3)) \cdot y) = (x \cdot ((x_0 x_1), x_2, x_3)) \cdot y - t_4(x_0, ..., x_3; l(4), v_2(4)) \cdot ((x \cdot (x_0, (x_1 x_2), x_3)) \cdot y) + t_4(x_0, ..., x_3; l(4), v_3(4); id)((x \cdot (x_0, x_1, (x_2 x_3))) \cdot y)$.

Define a $\mathcal{T}$-linear homomorphism $\mathbf{s}_n : K_n \to K_{n+1}$, which, for basic elements, has the form

(11) $\mathbf{s}_n(x_0, ..., x_{n+1}) = (1, x_0, ..., x_{n+1})$ for every $x_0, ..., x_{n+1}$ in G. From Formulas (9) and (10) in Definition 1 and (1) in Lemma 1 the identities

(12) $t_n(x_1, ..., x_n; q(n), u(n)|v) t_n(x_1, ..., x_n; u(n), q(n)|v^{-1}) = 1$

(13) $t_n(x_1, ..., x_n; q(n), u(n)) t_n(x_1, ..., x_n; u(n), w(n)) = t_n(x_1, ..., x_n; q(n), w(n))$ follow for every element $x_1,...,x_n$ in metagroup G. Vectors $q(n)$, $u(n)$, and $w(n)$ indicate the orders of their multiplication, $v \in S_n$ and $n \in \mathbf{N}$. The following identity is evident:

(14) $t_{n+1}(1, x_1, ..., x_n; q(n+1), u(n+1)|v(n+1)) = t_n(x_1, ..., x_n; q(n), u(n)|v(n))$ for data $q(n)$, $u(n)$ and $v(n)$ obtained from $q(n+1)$, $u(n+1)$, and $v(n+1)$ correspondingly by taking the identity $1b = b1 = b$ into account for each $b \in G$. Hence, $\mathbf{s}_n((x_0, ..., x_{n+1}) \cdot y) = (\mathbf{s}_n(x_0, ..., x_{n+1})) \cdot y$ for every $x_0, ..., x_{n+1}, y$ in G.

Let $p_n : K_{n+1} \to K_n$ be a $\mathcal{T}$-linear mapping, such that

(15) $p_n(a \otimes b) = a \cdot b$ and $p_n(b \otimes a) = b \cdot a$ for each $a \in K_n$ and $b \in A$. Therefore, from Formulas (13) and (14), we deduce that $p_n \mathbf{s}_n = id$ is the identity on K_n. Consequently, $\mathbf{s}_n$ is a monomorphism.

Therefore, from Formulas (1), (11), (13), and (14) we infer that $(\partial_{n+1}\mathbf{s}_n + \mathbf{s}_{n-1}\partial_n)(x_0, ..., x_{n+1}) = = \partial_{n+1}(1, x_0, ..., x_{n+1}) + \mathbf{s}_{n-1}(\sum_{j=0}^{n}(-1)^j t_{n+2}(x_0, ..., x_{n+1}; l(n+2), v_{j+1}(n+2)) \cdot < x_0, ..., x_{n+1} >_{j+1,n+2})) = = \sum_{j=0}^{n+1}(-1)^j t_{n+3}(1, x_0, ..., x_{n+1}; l(n+3), v_{j+1}(n+3)) \cdot < 1, x_0, x_1, ..., x_{n+1} >_{j+1,n+3} + + \sum_{j=0}^{n}(-1)^j t_{n+2}(x_0, ..., x_{n+1}; l(n+2), v_{j+1}(n+2)) \cdot < 1, x_0, ..., x_{n+1} >_{j+2,n+3} = (x_0, ..., x_{n+1})$, for every $x_0,...,x_{n+1}$ in G (see also Definitions 2 and 3 and the notations above).

Thus, the homotopy conditions

(16) $\partial_{n+1}\mathbf{s}_n + \mathbf{s}_{n-1}\partial_n = 1$ for each $n \geq 0$ are fulfilled, where 1 denotes the identity operator on K_n. Therefore, the recurrence relation

(17) $\partial_n \partial_{n+1} \mathbf{s}_n = \mathbf{s}_{n-2}\partial_{n-1}\partial_n$ is accomplished, since

$$\partial_n \partial_{n+1} \mathbf{s}_n = \partial_n(1 - \mathbf{s}_{n-1}\partial_n) = \partial_n - (\partial_n \mathbf{s}_{n-1})\partial_n = \partial_n - (1 - \mathbf{s}_{n-2}\partial_{n-1})\partial_n.$$

On the other hand, from Formula (11), it follows that K_{n+1}, as the left A-module, is generated by $\mathbf{s}_n K_n$. Then, proceeding by induction in n with the help of (17), we deduce that $\partial_n \partial_{n+1} = 0$ for each $n \geq 0$, since $\partial_0 \partial_1 = 0$ according to Formulas (1) and (5).

An opposite algebra A^{op} exists. The latter, as an $\mathcal{T}$-linear space, is the same, but has the multiplication $x \circ y = yx$ for each $x, y \in A^{op}$. Let $A^e := A \otimes_{\mathcal{T}} A^{op}$ denote the enveloping algebra of A. Apparently, $K_0 = A \otimes_{\mathcal{T}} A$ coincides with $A \otimes_{\mathcal{T}} A^{op}$ as a left and right A-module. Hence, the mapping $\partial_0 : K_0 \to K_{-1}$ provides the augmentation $\epsilon : A^e \to A$.

Thus, identity (16) means that the left complex $\mathcal{K}$ $0 \leftarrow A \underset{\partial_0}{\leftarrow} K_0 \underset{\partial_1}{\leftarrow} K_1 \underset{\partial_2}{\leftarrow} K_2 \leftarrow .. \underset{\partial_n}{\leftarrow} K_n \underset{\partial_{n+1}}{\longleftarrow} K_{n+1} \leftarrow ....$ is acyclic. □

Example 3. *For the Cayley–Dickson algebra A_n over a field F of characteristics not equal to two, let $G = G_n$, as the (multiplicative) metagroup, consist of all elements bi_k with $b \in \mathbf{\Psi}$, $k = 0, 1, 2, ...$, where $i_0, i_1, i_2, ...$ are generators of the Cayley–Dickson algebra A_n, $2 \leq n \leq \infty$. Then, $M = A_n^j$ is the module over $\mathbf{Z}[G]$, where $j \in \mathbf{N}$.*

Example 4. *For a topological space U, it is possible to consider the module $M = C(U, A_n^j)$ of all continuous mappings from U into A_n^j, $j \in \mathbf{N}$, A_n^j, which is supplied with the box product topology.*

Example 5. *If $(U, \mathcal{B}, \mu)$ is a measure space, where $\mu : \mathcal{B} \to [0, \infty)$ is a σ-additive measure on a σ-algebra $\mathcal{B}$ of a set U, for $\mathbf{F} = \mathbf{R}$ and $f_k = 1$ for each k, it is possible to consider the space $L_p((U, \mathcal{B}, \mu), A_n^j)$ of all L_p*

mappings from U into A_n^j, where A_n is taken relative to its norm induced by the scalar product $Re(\bar{y}z) = (y,z)$, $j \in \mathbf{N}$, $1 \le p \le \infty$.

Example 6. *For an additive group H, one can consider the trivial action of A on H. Therefore, the direct product $M \bigotimes H$ becomes an A-module for an A-module M. In particular, H may be a ring.*

Example 7. *If there is another ring $\mathcal{S}$ and a homomorphism $\phi : \mathcal{S} \to \mathcal{T}$, then each left (or right) $\mathcal{T}$-module M can be considered as a left (or right, correspondingly) $\mathcal{S}$-module by the rule $bm = (\phi b)m$ (or $mb = m(\phi b)$ correspondingly) for each $b \in \mathcal{S}$ and $m \in M$.*

Vice versa, if M is a right (or left) $\mathcal{S}$-module, then the right (or left, correspondingly) module exists $M_{(\phi)} = M \otimes_{\mathcal{S}} \mathcal{T}$ (or ${}_{(\phi)}M = \mathcal{T} \otimes_{\mathcal{S}} M$, correspondingly), which is called the right (or left correspondingly) covariant ϕ-extension of M. Similarly, the contravariant right and left extensions $M^{(\phi)} = Hom_{\mathcal{S}}(\mathcal{T}, M)$ or ${}^{(\phi)}M$ are defined for right or left $\mathcal{S}$-modules M, respectively.

This also can be applied to a metagroup algebra $A = \mathcal{S}[G]$ over a commutative associative unital ring $\mathcal{S}$ as in Example 1. Then, by changing a ring, we get right $A_{(\phi)}$ or $A^{(\phi)}$ and left ${}_{(\phi)}A$ or ${}^{(\phi)}A$ algebras over $\mathcal{T}$. Then, imposing the relation $ta = at$ for each $a \in A$ and $t \in T$ provides a metagroup algebra over $\mathcal{T}$, which also has a two-sided $\mathcal{T}$-module structure. It will be denoted by ${}_{(\phi)}A_{(\phi)}$ or ${}^{(\phi)}A^{(\phi)}$, respectively. Particularly, this is applicable to cases when $\mathbf{Z}[\mathbf{\Psi}] \subset \mathcal{S}$ or ϕ is an embedding.

Notation 1. *Let $A = \mathcal{T}[G]$ be a metagroup algebra (see Definition 2). Put $L_0 = \mathcal{T}$, $L_1 = A$ and by induction, $L_{n+1} = L_n \otimes_{\mathcal{T}} A$ for each natural number n.*

If N is a two-sided A-module, it can also be considered as a left A^e-module by the rule $(x \otimes y^*)b := (x \otimes b) \otimes y$ for each $x \in A$, $y^* \in A^{op}$, and $b \in N$, $A^e = A \otimes_{\mathcal{T}} A^{op}$ is an enveloping algebra, where A^{op} denotes the opposite algebra of A, where y^* in A^{op} corresponds to y in element A.

Theorem 1. *If $\mathcal{K}$ is an acyclic left A-complex for a metagroup algebra $A = \mathcal{T}[G]$, as in Proposition 1, and M is a two-sided A-module satisfying Conditions $(1-3)$ in Definition 3, then a co-chain complex $Hom(\mathcal{L}, M)$ exists:*

(1) $0 \to Hom_{\mathcal{T}}(L_0, M) \underset{\epsilon^*}{\to} Hom_{\mathcal{T}}(L_1, M) \underset{\delta^1}{\to} Hom_{\mathcal{T}}(L_2, M) \underset{\delta^2}{\to} Hom_{\mathcal{T}}(L_3, M) \underset{\delta^3}{\to} Hom_{\mathcal{T}}(L_4, M) \underset{\delta^4}{\to} \dots$

such that $f \in Hom_{\mathcal{T}}(L_1, M)$ is a co-cycle, if and only if f is a $\mathcal{T}$-linear derivation from A into M.

Proof. Notation 1 and Example 2 permit each basic element $(x_0, ..., x_{n+1})$ of K_n over $\mathcal{T}$ to be written as

(1) $(x_0, ..., x_{n+1}) = t_{n+2}(x_0, ..., x_{n+1}; l(n+2), w(n+2)) \cdot ((x_0 \otimes (x_1, ..., x_n)) \otimes x_{n+1})$ and

(2) $(x_0, ..., x_{n+1}) = t_{n+2}(x_0, ..., x_{n+1}; l(n+2), w(n+2)) \cdot (z \otimes (x_1, ..., x_n))$, where $(x_1, ..., x_n)$ is a basic element in L_n for every $x_0, ..., x_{n+1}$ in G, $\{x_0, ..., x_{n+1}\}_{w(n+2)} = (x_0\{x_1, ..., x_n\}_{l(n)})x_{n+1}$, $z \in A^e$, $z = x_0 \otimes x_{n+1}^*$.

Each homomorphism $f \in Hom_{\mathcal{T}}(L_n, M)$ is characterized by its values on elements $(x_1, ..., x_n)$, where x_1,..,x_n belong to a metagroup G. Consider f as a $\mathcal{T}$-linear function from A^n into M. Since M satisfies Conditions $(1-3)$ in Definition 3, then f has the decomposition

(2) $f(x_1, ..., x_n) = \sum_{g \in G} f_g(x_1, ..., x_n)$,

where $f_g : G^n \to M_g$ for every g and $x_1, ..., x_n$ in G.

Therefore, the restrictions follow from Conditions $(1-3)$ in Definition 3, which take into account the nonassociativity of G:

(3) $(xy) \cdot f_g(x_1, ..., x_n) = t_3(x, y, g) \cdot (x \cdot (y \cdot f_g(x_1, ..., x_n)))$,

(4) $t_3(g, x, y) \cdot (f_g(x_1, ..., x_n) \cdot (xy)) = (f_g(x_1, ..., x_n) \cdot x) \cdot y$,

(5) $(x \cdot f_g(x_1, ..., x_n)) \cdot y = t_3(x, g, y) \cdot (x \cdot (f_g(x_1, ..., x_n) \cdot y))$
for every g and $x, y, x_1, ..., x_n$ in G, where coefficients t_3 are prescribed by Formula (9) in Definition 1. Also,

(6) $x \cdot f_g(x_1, ..., x_n) := x \cdot (f_g(x_1, ..., x_n))$ and

(7) $f_g(x_1, ..., x_n) \cdot y := (f_g(x_1, ..., x_n)) \cdot y$.

For $n = 0$ and $g = e$, naturally, the identities are fulfilled:

(8) $(xy) \cdot f_e(\,) = x \cdot (y \cdot f_e(\,))$, $(f_e(\,) \cdot x) \cdot y = f_e(\,) \cdot (xy)$ and $(x \cdot f_e(\,)) \cdot y = x \cdot (f_e(\,) \cdot y)$.

A co-boundary operator exists that takes into account the nonassociativity of the (multiplicative) metagroup G:

(9) $(\delta^n f)(x_1, ..., x_{n+1}) = \sum_{j=0}^{n+1} (-1)^j t_{n+1}(x_1, ..., x_{n+1}; l(n+1), u_{j+1}(n+1)) \cdot [f, x_1, x_2, ..., x_{n+1}]_{j+1,n+1}$,
where

(10) $[f, x_1, ..., x_{n+1}]_{1,n+1} := x_1 \cdot f(x_2, ..., x_{n+1})$, $\{x_1, ..., x_{n+1}\}_{u_1(n+1)} = x_1\{x_2, ..., x_{n+1}\}_{l(n)}$;

(11) $[f, x_1, ..., x_{n+1}]_{2,n+1} := f((x_1x_2), ..., x_{n+1})$, $\{x_1, ..., x_{n+1}\}_{u_2(n+1)} = \{(x_1x_2), ..., x_{n+1}\}_{l(n)}$;...;

(12) $[f, x_1, ..., x_{n+1}]_{n+1,n+1} := f(x_1, x_2, ..., (x_nx_{n+1}))$; $\{x_1, ..., x_{n+1}\}_{u_{n+1}(n+1)} = \{x_1, ..., (x_nx_{n+1})\}_{l(n)}$;

(13) $[f, x_1, ..., x_{n+1}]_{n+2,n+1} := f(x_1, x_2, ..., x_n) \cdot x_{n+1}$, $\{x_1, ..., x_{n+1}\}_{u_{n+2}(n+1)} = \{x_1, ..., x_{n+1}\}_{l(n+1)} = (...((x_1x_2)x_3)...x_n)x_{n+1}$; with $u_0(n+1) = l(n+1)$.

From G^{n+1} onto K_{n+1}, the homomorphism $(\delta^n f)$ is extended by the $\mathcal{T}$-linearity. On the other hand, Condition (1) in Definition 3 implies that

(14) For each $b \in G$, $h_{1,b}$ exists, so that $h_{1,b} : K_{n+1} \to M_1$ and $f_b = h_{1,b}L_b$, where L_b is the left multiplication operator on b:

(15) $(h_{1,b}L_b)(x_1, ..., x_n) = b \cdot (h_{1,b}(x_1, ..., x_n))$ for every $x_1, ..., x_n$ in G. Moreover, $zg = 0$ (or $gz = 0$) in $\mathbf{Z}[G]$ for $g \in G$ and $z \in \mathbf{Z}[G]$, if and only if $z = 0$, since G is a metagroup.

By virtue of Proposition 1, these formulas imply that $\delta^{n+1} \circ \delta^n = 0$ for each n, since $(\delta^{n+1} \circ \delta^n f)(x_1, ..., x_{n+2}) = f(\partial_{n-1} \circ \partial_n(x_1, ..., x_{n+2}))$ for every $x_1, ..., x_{n+2}$ in G. Thus, the complex given by formula (1) is exact.

Particularly, $f \in Hom_{\mathcal{T}}(L_0, M)$ is a co-cycle if and only if

(16) $(\delta^0 f)(x) = xf(\,) - f(\,)x = 0$ for each $x \in G$.

We mention that $Hom_{\mathcal{T}}(L_0, M)$ is isomorphic with M.

The one-dimensional co-chain $f \in Hom_{\mathcal{T}}(L_1, M)$ is determined by the mapping $f : G \to M$. Taking Formula (9) into account, we infer that it is a co-cycle if and only if

(17) $t_2(x, y; l(2), u_1(2)) \cdot x \cdot f(y) - t_2(x, y; l(2), u_2(2)) \cdot f(xy) + t_2(x, y; l(2), u_3(2)) \cdot f(x) \cdot y = x \cdot f(y) - f(xy) + f(x) \cdot y = 0$

for each x and y in G. That is, f is a derivation from the metagroup G into the G-module M. There is the embedding $\mathcal{T} \hookrightarrow A$ of $\mathcal{T}$ into A as $\mathcal{T}e$, since $e = 1 \in G$. Thus, f has a $\mathcal{T}$-linear extension to a $\mathcal{T}$-linear derivation from A into M by the following formula:

(18) $f(xy) = x \cdot f(y) + f(x) \cdot y$.

□

Remark 1. *Suppose that the conditions of Theorem 1 are fulfilled. A two-dimensional co-chain is a 2-co-cycle, if and only if* $(\delta^2 f)(x_1, x_2, x_3) = \sum_{j=0}^{3} (-1)^j t_3(x_1, x_2, x_3; l(3), u_{j+1}(3)) \cdot [f, x_1, x_2, x_3]_{j+1,3} = t_3(x_1, x_2, x_3; l(3), u_1(3)) \cdot x_1 \cdot f(x_2, x_3) - t_3(x_1, x_2, x_3; l(3), u_2(3)) \cdot f((x_1x_2), x_3) + t_3(x_1, x_2, x_3; l(3), u_3(3)) \cdot f(x_1, (x_2x_3)) - t_3(x_1, x_2, x_3; l(3), u_4(3)) \cdot f(x_1, x_2) \cdot x_3 = 0$.

That is,

(1) $t_3(x_1, x_2, x_3) \cdot x_1 \cdot f(x_2, x_3) + t_3(x_1, x_2, x_3) \cdot f(x_1, (x_2x_3)) = f((x_1x_2), x_3) + f(x_1, x_2) \cdot x_3$

for each x_1, x_2 *and* x_3 *in* G.

Usually, $Z^n(A,M)$ denotes the set of all n-co-cycles, and the notation $B^n(A,M)$ is used for the set of n-co-boundaries in $H_{\mathcal{T}}(L_n, M)$. Since, as the additive group, M is commutative, then there are defined groups of cohomologies $H^n(A,M) = Z^n(A,M)/B^n(A,M)$ as the quotient (additive) groups.

For $n = 0$, the co-boundaries are set as zero, and hence, $H^0(A,M) \cong M^A$. In the case $n = 1$, a mapping $f : A \to M$ is a co-boundary if the element $m = h(\ ) \in M$ exists, for which $f(x) = xm - mx$ for each $x \in A$. Such a derivation f is called an inner derivation of A defined by an element $m \in M$. The set of all inner derivations is denoted by $Inn_{\mathcal{T}}(A,M)$.

From the cohomological point of view, the additive group $H^1(A,M)$ is interpreted as the group of all outer derivations $H^1(A,M) \cong Out_{\mathcal{T}}(A,M) \cong Der_{\mathcal{T}}(A,M)/Inn_{\mathcal{T}}(A,M)$, where $Z^1(A,M) = Der_{\mathcal{T}}(A,M)$; $Inn_{\mathcal{T}}(A,M) = B^1(A,M)$, where the family of all derivations ($\mathcal{T}$-homogeneous derivations) from X into a two-sided module M over $\mathcal{T}$ is denoted by $Der(X,M)$ (or $Der_{\mathcal{T}}(X,M)$ respectively).

A two-co-chain $f : G \times G \to M$ is a two-co-boundary, if a one-co-chain $h : G \to M$ exists such that for each x and y in G, the following identity is fulfilled:

(2) $f(x,y) = (\delta h)(x,y) = \sum_{j=0}^{2}(-1)^j t_2(x,y;l(2),u_{j+1}(2)) \cdot [h, x_1, x_2]_{j+1,2},$
$= x \cdot h(y) - h(xy) + h(x) \cdot y.$

Let $A = \mathcal{T}[G]$ be a metagroup algebra over a commutative associative unital ring $\mathcal{T}$ (also see Definitions 1–3).

Let M, N, and P be left A-modules, and a short exact sequence exists:

(3) $0 \to M \underset{\xi}{\to} P \underset{\eta}{\to} N \to 0,$

where ξ is an embedding, such that ξ and η are left A-homomorphisms. Then, P is called an enlargement of a left A-module M with the help of a left A-module N. If there is another enlargement of M with the help of N,

(4) $0 \to M \underset{\xi'}{\to} P' \underset{\eta'}{\to} N \to 0$

such that an isomorphism $\pi : P \to P'$ exists for which $\pi\xi = \xi' 1_M$ and $1_N \eta = \eta'\pi$, then enlargements (3) and (4) are called equivalent, where $1_M : M \to M$ notates the identity mapping, $1_M(m) = m$ for each $m \in M$.

It is said that an enlargement clefts, if and only if a left A-homomorphism $\omega : N \to P$ exists, fulfilling the restriction $\eta\omega = 1_N$.

In the particular case when $P = M \oplus N$, ξ is also an identifying mapping with the first direct summand, and η is a projection on the second direct summand, an enlargement is called trivial.

Theorem 2. *Let A be a nonassociative metagroup algebra and let M and N be left A-modules, where $A = \mathcal{T}[G]$, G is a metagroup (see Definitions 2 and 3). Then, the family $T = Hom_{\mathcal{T}}(N,M)$ can be supplied with a two-sided A-module structure, such that $H^1(A,T)$ is the set of classes of modules M with the quotient module N.*

Proof. The family $T = Hom_{\mathcal{T}}(N,M)$ evidently has the structure of a left module over a ring $\mathcal{T}$ (see Definition 3), and it can be supplied with a two-sided A-module structure:

(1) $\forall\, r \in T$ and $\forall\, n \in N$ and $\forall\, a \in A$:
$(a \cdot r)(n) = a \cdot (r(n))$ and $(r \cdot a)(n) = r(a \cdot n)$.

By virtue of Theorem 1, each element $f \in Z^1(A,T)$ induces a (generalized) derivation by Formula (18) in Theorem 1. Each zero-dimensional co-chain $m \in M$ provides an inner derivation $\delta^0 m(a) = am - ma$ due to formula (16) in Theorem 1. Then, one co-cycle f induces an enlargement by Formula (3) in Remark 1 with $P = M \oplus N$ being the direct sum of left A-modules in which N is a submodule and with the left action of A on N: $a \circ n = a \cdot n + f(a) \cdot n$ for each $n \in N$ and $a \in A$.

Suppose that a class of an one co-cycle f is zero, an element $u \in T$ exists so that $f = \delta^1 u$. Then, elements of the form $m + u(m)$ form its submodule M', which is isomorphic with M. Moreover, $P = N \oplus M'$ is the direct sum of A-modules. Thus, an enlargement is trivial.

Vice versa, suppose that an enlargement given by formula (3) in Remark 1 exists. That is, a left A-homomorphism $\gamma : N \to P$ satisfying the restriction $\eta\gamma = 1_N$ exists, where 1_N notates the identity mapping on N. It induces $f \in Z^1(A, T)$, such that $f(a)n = \gamma(a \cdot n) - (a \cdot \gamma)(n)$ for all $n \in N$ and for each element a of the algebra A.

Suppose that there is another enlargement which clefts, that is, a left $\mathcal{T}$-homomorphism $\omega : N \to P$ exists, fulfilling the restriction $\eta\omega = 1_N$. We put $u(n) = \omega(n) - \gamma(n)$ for each $n \in N$; hence, $u \in T$. Then, $f_1 = \delta^1 u$ is a co-cycle of zero class. □

Theorem 3. *Suppose that A is a nonassociative metagroup algebra over a commutative associative unital ring $\mathcal{T}$, a left A-module N and a two-sided A-module M are given. Then, $H^2(A, T)$ is the set of classes of enlargements of A with a kernel M such that $M^2 = \{0\}$ and with the quotient algebra A. Moreover, an action of A on M in this enlargement coincides with the structure of a two-sided A-module on M.*

Proof. If P is an enlargement with a kernel M such that $M^2 = 0$ and a quotient module $A = P/M$ and $a = p + M$ with $p \in P$, then $a \cdot m = p \cdot m$ and $m \cdot a = m \cdot p$ supply M with the two-sided A-module structure. Take a $\mathcal{T}$-linear mapping $\gamma : A \to P$ inverse from the left to a natural epimorphism and put $f(a,b) = \gamma(ab) - \gamma(a)\gamma(b)$ for each a and b in A. Then, we infer that $\gamma(a(bc)) = f(a,bc) + \gamma(a)\gamma(bc) = f(a,bc) + \gamma(a)(f(b,c) + \gamma(b)\gamma(c))$ and $\gamma((ab)c) = f(ab,c) + \gamma(ab)\gamma(c) = f(ab,c) + (f(a,b) + \gamma(a)\gamma(b))\gamma(c)$, consequently, $0 = t_3(a,b,c)\gamma(a(bc)) - \gamma((ab)c) = t_3(a,b,c)f(a,bc) + t_3(a,b,c)\gamma(a)(f(b,c) + \gamma(b)\gamma(c)) - f(ab,c) - (f(a,b) + \gamma(a)\gamma(b))\gamma(c)$.

Taking into account that $\gamma(a)m = a \cdot m$ and $m\gamma(a) = m \cdot a$ for each $m \in M$ and $a \in A$, we deduce using Formula (1) in Remark 1 that $0 = t_3(a,b,c) \cdot a \cdot f(b,c) - f(ab,c) + t_3(a,b,c) \cdot f(a,bc) - f(a,b) \cdot c = (\delta^2 f)(a,b,c)$.

Thus, $f \in B^2(A, M)$ and hence, $f = (\delta^1 h)$ with $h \in C^1(A, M) := Hom_{\mathcal{T}}(A, M)$.

It remains to prove that the set S of all elements $\gamma(a) + h(a)$ forms a subalgebra isomorphic with A in P . From the construction of S, it follows that S is a a two-sided $\mathcal{T}$-module. We verify that it is closed relative to the multiplication for all a and b in A:

$(\gamma(a) + h(a))(\gamma(b) + h(b)) = \gamma(a)\gamma(b) + \gamma(a)h(b) + h(a)\gamma(b) = \gamma(ab) - f(a,b) + ah(b) + h(a)b = \gamma(ab) + h(ab) + ah(b) - h(ab) + h(a)b - f(a,b) = \gamma(ab) + h(ab) + (\delta^1 h)(a,b) - f(a,b) = \gamma(ab) + h(ab)$.

If A, M, and f are given, then an enlargement P can be constructed as the direct sum $P = M \oplus A$ of two-sided $\mathcal{T}$-modules and with the multiplication rule $(m_1 + b_1)(m_2 + b_2) = m_1 b_2 + m_2 b_1 + f(b_1, b_2) + b_1 b_2$ for every m_1 and m_2 in M and b_1 and b_2 in A. It rests to verify that this multiplication rule is homogeneous over $\mathcal{T}$ and right and left distributive. At first, we evidently get $(m_1 + b_1)(s(m_2 + b_2)) = (s(m_1 + b_1))(m_2 + b_2) = s((m_1 + b_1)(m_2 + b_2)) = sm_1 b_2 + sm_2 b_1 + sf(b_1, b_2) + sb_1 b_2$ and $(sp)(m_1 + b_1) = s(p(m_1 + b_1))$ for all $s, p \in \mathcal{T}$ and m_1 and m_2 in M and b_1 and b_2 in A, since $\mathcal{T} \subset \mathcal{C}(A)$ and $f(s,p) = 0$. Moreover, we infer that $(m_1 + b_1)((m_2 + b_2) + (m_3 + b_3)) = (m_1 + b_1)((m_2 + m_3) + (b_2 + b_3)) = m_1(b_2 + b_3) + (m_2 + m_3)b_1 + f(b_1, b_2 + b_3) + b_1(b_2 + b_3) = m_1 b_2 + m_1 b_3 + m_2 b_1 + m_3 b_1 + f(b_1, b_2) + f(b_1, b_3) + b_1 b_2 + b_1 b_3 = (m_1 + b_1)(m_2 + b_2) + (m_1 + b_1)(m_3 + b_3)$, and analogously, $((m_1 + b_1) + (m_2 + b_2))(m_3 + b_3) = (m_1 + b_1)(m_3 + b_3) + (m_2 + b_2)(m_3 + b_3)$ for all m_1, m_2 and m_3 in M and b_1, b_2 and b_3 in A. □

Definition 4. *Let M and P and N be two-sided A-modules, where A is a nonassociative metagroup algebra over a commutative associative unital ring $\mathcal{T}$. An A-homomorphism (isomorphism) $f : M \to P$ is called a right (operator) A-homomorphism (isomorphism) if it is such for M and N as right A-modules, that is, $f(x+y) = f(x) + f(y)$ and $f(xa) = f(x)a$ for each x and y in M and $a \in A$ (see also Definition 3). If an algebra A is specified, a homomorphism (isomorphism) may be written for short instead of an A-homomorphism (an A-isomorphism, respectively).*

An enlargement (P, η) of M by N is called right inessential if a right isomorphism $\gamma : N \to P$ exists, satisfying the restriction $\eta\gamma|_N = 1|_N$.

Theorem 4. *Suppose that M is a two-sided A-module, where A is a nonassociative metagroup algebra over a commutative associative unital ring $\mathcal{T}$. Then, for each $n \geq 0$, a two-sided A-module P_n exists such that $H^{n+1}(A, M)$ is isomorphic with the additive group of equivalence classes of right inessential enlargements of M by P_n.*

Proof. Consider two right inessential enlargements (E_1, η_1) and (E_2, η_2) of M by N, where ξ_1 and ξ_2 are embeddings of M into E_1 and E_2 correspondingly. Take a submodule Q of $E_1 \oplus E_2$ consisting of all elements (x_1, x_2) satisfying the condition $\eta_1(x_1) = \eta_2(x_2)$. Then, a quotient module Q/T exists, where $T = \{(\xi m, -\xi m) : m \in M\}$. Therefore, $(\xi_1 M \oplus \xi_2 M)/T$ is isomorphic with M, and homomorphisms η_1 and η_2 induce a homomorphism η of Q/T onto N. Hence, the submodule $ker(\eta)$ is isomorphic with M. Then, an addition of enlargements is prescribed by the formula $(E_1, \eta_1) + (E_2, \eta_2) := (Q/T, \eta)$. Evidently, sums of equivalent enlargements are equivalent.

For an enlargement (E, η) of M by N, one takes the direct sum of modules $E \oplus M$ and puts T_b to be its submodule consisting of all elements $(\xi m, -b\xi m)$ with $m \in M$, where ξ is an embedding of M into E, $b \in \mathcal{T}$. Therefore, a homomorphism η induces a homomorphism ${}_b\eta$ of $(E \oplus M)/T_b$ onto N, since the mapping $(\xi m, m) \mapsto b\xi m + m$ is a homomorphism of $(\xi M) \oplus M$ onto M. Also, the ring $\mathcal{T}$ is commutative and associative. This induces an enlargement of M by N, denoted by $({}_bE, {}_b\eta)$ and hence, an operation of scalar multiplication of an enlargement (E, η) on $b \in \mathcal{T}$. From this construction, it follows that equivalent enlargements have equivalent scalar multipliers on $b \in \mathcal{T}$.

Let P_n be a $\mathcal{T}$-linear span of all elements $(x_1, ..., x_{n+1})$ with $x_1,...,x_{n+1}$ in G such that $((bx_1), x_2, ..., x_{n+1}) = (x_1, ..., (bx_{n+1}))$ for each $b \in \mathcal{T}$. Next, we put

(1) $(x_1, ..., x_{n+1}) \cdot y := t_{n+2}(x_1, ..., x_{n+1}, y; l(n+2), u_{n+2}(n+2)) \cdot (x_1, ..., x_n, (x_{n+1}y))$ and

(2) $y \cdot (x_1, ..., x_{n+1}) = \sum_{j=1}^{n+1} (-1)^{j+1} \cdot$

$$t_{n+2}(y, x_1, ..., x_{n+1}; u_1(n+2), u_{j+1}(n+2)) \cdot < y, x_1, x_2, ..., x_{n+1} >_{j,n+2}$$

(also see Notations (2–4) of Proposition 1 and (10)–(13) in Theorem 1) for every $y, x_1, ..., x_{n+1}$ in G. That is, P_n is the two-sided A-module, where A has the unit element.

By $R_n = R(P_n, M)$, we denote the family of all right homomorphisms of P_n into M. For each $p \in R_n$, let an arbitrary element $\dot{p} \in C^n(A, M)$ in the additive group of all n co-chains (that is, n times $\mathcal{T}$-linear mappings of A into M) on A with values in M be prescribed by the formula $\dot{p}(a_1, ..., a_n) = p(a_1, ..., a_n, 1)$ for all $a_1, ..., a_n$ in A. Consequently, $(\dot{p}(a_1, ..., a_n)) \cdot y = p(a_1, ..., a_n, y)$ for each $y \in A$, since $t_{n+3}(x_1, ..., x_{n+1}, 1, g; l(n+3), u_{n+3}(n+3)) = 1$ for all $x_1, ..., x_{n+1}$ and g in G. This makes the mapping $p \mapsto \dot{p}$ an $\mathcal{T}$-linear isomorphism of R_n onto $C^n(A, M)$.

Supply $C^n(A, M)$ with a two-sided A-module structure

(3) $(x_0 \cdot f)(x_1, ..., x_n) = x_0 \cdot (f(x_1, ..., x_n))$ and

(4) $(f \cdot x_0)(x_1, .., x_n) = \sum_{k=0}^{n-1} (-1)^k t_{n+1}(x_0, x_1, ..., x_n; u_1(n+1), u_{k+2}(n+1)) \cdot f(x_0, ..., x_k x_{k+1}, ..., x_n)$ $+(-1)^n (f(x_0, ..., x_{n-1})) \cdot x_n$

for each $f \in C^n(A, M)$ and all $x_0, x_1, ..., x_n$ in G, extending f by $\mathcal{T}$-linearity on A from G, where $u_j(n+1)$ are given by Formulas (10)–(13) in Theorem 1. Thus, the mapping $p \mapsto \dot{p}$ is an operator isomorphism. Consequently, $H^p(A, R_n)$ is isomorphic with $H^p(A, C^n(A, M))$ for each integer (n and p) such that $n \geq 0$ and $p \geq 0$. On the other hand, $H^p(A, C^n(A, M))$ is isomorphic with $H^{p+n}(A, M)$ for each $p \geq 1$; hence, $H^p(A, R_n)$ is isomorphic with $H^{p+n}(A, M)$.

By virtue of Theorem 2 applied with $p = 1$, we infer that $H^{n+1}(A, M)$ is isomorphic with the additive group of equivalence classes of right inessential enlargements of M by P_n. □

Theorem 5. *Let M be a two-sided A-module, where A is a nonassociative metagroup algebra over a commutative associative unital ring $\mathcal{T}$. Then, to each $n+1$-co-cycle $f \in Z^{n+1}(A,M)$, an enlargement of M by a two-sided A-module P_n corresponds such that f becomes a co-boundary in it.*

Proof. An $n+1$-co-cycle $f \in Z^{n+1}(A,M)$ induces an enlargement (E,η) of M by P_n due to Theorem 4. An element h in $Z^1(A,R_n)$ corresponding to f is characterized by the equality

$$(h(x_1))(x_2,...,x_{n+1},1) = t_{n+1}(x_1,...,x_{n+1};u_1(n+1),l(n+1)) \cdot f(x_1,...,x_{n+1})$$

for all $x_1,...,x_{n+1}$ in G. This enlargement (E,η) as the two-sided A-module is $P_n \oplus M$ such that $x_1 \cdot ((x_2,...,x_{n+1},1),0) = (x_1 \cdot (x_2,...,x_{n+1}), f(x_1,...,x_{n+1}))$. Let $\gamma(a_1,...,a_n) = (a_1,...,a_n,0)$ for all $a_1,...,a_n$ in A. Therefore, we deduce that $f(x_1,...,x_{n+1}) = t_{n+1}(x_1,...,x_{n+1};l(n+1),u_1(n+1)) \cdot \{x_1 \cdot \gamma(x_2,...,x_{n+1},1) - \gamma(x_1 \cdot (x_2,...,x_{n+1},1))\}$.

An n-co-chain $v \in C^n(A,E)$, defined by $v(a_1,...,a_n) = ((a_1,...,a_n,1),0)$, exists for all $a_1,...,a_n$ in A. Thus, $f = \delta v$. □

Theorem 6. *Let A be a nonassociative metagroup algebra over a commutative associative unital ring $\mathcal{T}$. Then, an algebra B over $\mathcal{T}$ exists such that B contains A and each $\mathcal{T}$-homogeneous derivation $d : A \to A$ is the restriction of an inner derivation of B.*

Proof. Naturally, an algebra A has the structure of a two-sided A-module. In view of Theorem 1, each derivation of the two-sided algebra A can be considered an element of $Z^1(A,A)$.

Applying Theorem 5 by induction, one obtains a two-sided A-module Q containing M for which an arbitrary element of $Z^{n+1}(A,M)$ is represented as the co-boundary of an element of $C^n(A,Q)$. At the same time, M and Q satisfy Conditions (1)–(3) in Definition 3. This implies that the natural injection of $H^{n+1}(A,M)$ into $H^{n+1}(A,Q)$ maps $H^{n+1}(A,M)$ into zero.

Therefore, a two-sided A-module E exists, which, as a two-sided $\mathcal{T}$-module, is a direct sum, $A \oplus P$, and P is such that for each $f \in Z^1(A,A)$, an element $p \in P$ exists that generally depends on f with the property $f(a) = a \cdot p - p \cdot a$. The metagroup G corresponds to the algebra A. By enlarging P, if necessary, we can consider that to P, a metagroup G also corresponds in such a manner that properties (1)–(3) in Definition 3 are fulfilled.

Now, we take $A \oplus P$ as the underlying two-sided $\mathcal{T}$-module of B and supply it with the multiplication $(a_1,p_1)(a_2,p_2) := (a_1a_2, a_1 \cdot p_2 + p_1 \cdot a_2)$ as the semidirect product for each a_1, a_2 in A and p_1, p_2 in P. An embedding ξ of A into B is $\xi(a) = (a,0)$ for each a in A. This implies that $f(a) = (a,0)(0,p) - (0,p)(a,0) = a(0,p) - (0,p)a$. □

Theorem 7. *Suppose that A is a nonassociative metagroup algebra of finite order over a commutative associative unital ring $\mathcal{T}$ and M is a finitely generated two-sided A-module. Then, M is semisimple if and only if its cohomology group is null $H^n(A,M) = 0$ for each natural number $n \geq 1$.*

Proof. Certainly, if E is an A-module and N is its A-submodule, then a natural quotient morphism $\pi : E \to E/N$ exists. Therefore, an enlargement (E,η) of a two-sided A-module M by a two-sided A-module N is inessential if and only if there is a submodule T in E complemented to $\xi(M)$ such that T is isomorphic with $E/\xi(M)$, where ξ is an embedding of M into E. If M is semisimple, then it is either a simple or a finite product of simple modules, since M is finitely generated. For a finitely generated module E and its submodule N, the quotient module E/N is not isomorphic with E, since the algebra A is of finite order over the commutative associative unital ring $\mathcal{T}$.

By virtue of Theorems 4 and 5, if, for an algebra A, its corresponding finitely generated two-sided A-modules are semisimple, then its cohomology groups of dimension $n \geq 1$ are zero.

Vice versa, suppose that $H^n(A,M) = 0$ for each natural number $n \geq 1$. Consider a finitely generated two-sided A-module E and its two-sided A-submodule N. At first, we take into account the right A-module structure E_r of E with the same right transformations, but with zero left transformations. Then, the left inessential (E_r,η_r) enlargement of M_r by $N_r = E_r/\xi_r(M_r)$ exists, where $\eta_r : E_r \to$

$E_r/\xi_r(M_r)$ is the quotient mapping and ξ_r is an embedding of M_r into E_r. From Theorem 4, it follows that the enlargement (E_r, η_r) is right inessential. Analogously, considering the left A-module structures E_l and M_l we infer that (E_l, η_l) is also left inessential. □

Note 2. *Let A be a nonassociative metagroup algebra over a commutative associative unital ring $\mathcal{T}$ with a characteristic char$(\mathcal{T})$ other than two and three. Its opposite algebra A^{op} exists. The latter, as an F-linear space, is the same, but with the multiplication $x \circ y = yx$ for each $x, y \in A^{op}$. To each element $h \in A$ or $y \in A^{op}$, there is posed a left multiplication operator L_h by the formula $xL_h = hx$ or a right multiplication operator $xR_y = xy$ for each $x \in A$, respectively. Having the anti-isomorphism operator $S : A \to A^{op}$, $A \ni x \mapsto xS \in A^{op}$, $S(xy) = S(y)S(x)$, we get*

(1) *$R_hS = SL_{hS}$ and $L_hS = SR_{hS}$ for each x and h in A. Then, taking into account (1) analogously to formula (4) in Theorem 4, we put $x_0 \cdot (L_{x_1}, R_{x_2}) = \mathrm{t}_3^{-1}(x_0, x_1, x_2) \cdot (x_0L_{x_1})L_{x_2S} - x_0(L_{x_1}R_{x_2})$ $+\mathrm{t}_3^{-1}(x_0, x_1, x_2) \cdot (x_0R_{x_1S})R_{x_2}$.*
Then, taking into account the multipliers t_3, this gives

(2) *$x_0 \cdot (L_{x_1}, R_{x_2}) = x_0(L_{x_1}L_{x_2S} - L_{x_1}R_{x_2} + R_{x_1S}R_{x_2})$ for all x_0, x_1, x_2 in G. Next, symmetrically, $S(x_0 \cdot (L_{x_1}, R_{x_2}))$ provides the formula for $(L_{y_1}, R_{y_2}) \cdot y_0$ for each y_0, y_1 and y_2 in G. We consider the enveloping algebra $A^e = A \otimes_{\mathcal{T}} A^{op}$. Extending these rules by $\mathcal{T}$-linearity on A and $A \otimes_{\mathcal{T}} A^{op}$ from G one supplies the tensor product $M = A^e$ over $\mathcal{T}$ with the two-sided A-module structure.*

Corollary 1. *Let A be a semisimple, nonassociative metagroup algebra of finite order over a commutative associative unital ring $\mathcal{T}$ with a characteristic char$(\mathcal{T})$ other than two and three, and let M be a two-sided A-module described in Note 2. Then, $H^n(A, M) = 0$ for each natural number $n \geq 1$.*

Proof. Since A is semisimple, then the module M from Note 2 is semisimple. Consequently, the statement of this corollary follows from Theorem 7. □

3. Products of Metagroups

The main subject of this paper are cohomologies on metagroups. Nonetheless, in this section, it is shortly demonstrated that there are abundant families of metagroups besides those which appear in areas described in the introduction.

Theorem 8. *Let G_j be a family of metagroups (see Definition 1 in Section 2), where $j \in J$, J is a set. Then, their direct product $G = \prod_{j \in J} G_j$ is a metagroup and*

(1) $C(G) = \prod_{j \in J} C(G_j)$.

Proof. Each element $a \in G$ is written as $a = \{a_j : \forall j \in J,\ a_j \in G_j\}$. Therefore, a product $ab = \{c : \forall j \in J,\ c_j = a_jb_j,\ a_j \in G_j,\ b_j \in G_j\}$ is a single-valued binary operation on G. Then, we get that $a \setminus b = \{d : \forall j \in J,\ d_j = a_j \setminus b_j,\ a_j \in G_j,\ b_j \in G_j\}$ and $a/b = \{d : \forall j \in J,\ d_j = a_j/b_j,\ a_j \in G_j,\ b_j \in G_j\}$. Moreover, $e_G = \{\forall j \in J,\ e_{G_j}\}$ is a neutral element in G, where e_{G_j} denotes a neutral element in G_j for each $j \in J$. Thus, Conditions (1)–(3) of Definition 1 in Section 2 are satisfied.

From Conditions (4)–(7) of Definition 1 in Section 2 for each G_j, we infer that

(2) $Com(G) := \{a \in G : \forall b \in G,\ ab = ba\} = \{a \in G :\ a = \{a_j : \forall j \in J, a_j \in G_j\}; \forall b \in G,\ b = \{b_j : \forall j \in J, b_j \in G_j\}; \forall j \in J,\ a_jb_j = b_ja_j\} = \prod_{j \in J} Com(G_j)$,
(3) $N_l(G) := \{a \in G :\ \forall b \in G,\ \forall c \in G,\ (ab)c = a(bc)\} = \{a \in G :\ a = \{a_j : \forall j \in J, a_j \in G_j\};\ \forall b \in G,\ b = \{b_j : \forall j \in J, b_j \in G_j\};\ \forall c \in G,\ c = \{c_j : \forall j \in J, c_j \in G_j\};\ \forall j \in J,\ (a_jb_j)c_j = a_j(b_jc_j)\} = \prod_{j \in J} N_l(G_j)$, and similarly,
(4) $N_m(G) = \prod_{j \in J} N_m(G_j)$ and
(5) $N_r(G) = \prod_{j \in J} N_r(G_j)$.

This and (8) of Definition 1 in Section 2 imply that

(6) $N(G) = \prod_{j\in J} N(G_j)$. Thus,
(7) $\mathcal{C}(G) := Com(G) \cap N(G) = \prod_{j\in J} \mathcal{C}(G_j)$.

Let a, b, and c be in G. Then,

$(ab)c = \{(a_j b_j)c_j : \forall j \in J,\ a_j \in G_j, b_j \in G_j, c_j \in G_j\} = \{t_{3,G_j}(a_j, b_j, c_j)a_j(b_j c_j) : \forall j \in J,\ a_j \in G_j, b_j \in G_j, c_j \in G_j\} = t_{3,G}(a,b,c)a(bc)$, where
(8) $t_{3,G}(a,b,c) = \{t_{3,G_j}(a_j, b_j, c_j) : \forall j \in J,\ a_j \in G_j, b_j \in G_j, c_j \in G_j\}$.

Therefore, Formulas (7) and (8) imply that Condition (9) of Definition 1 in Section 2 is also satisfied. Thus, G is a metagroup. □

Remark 2. (1) *Let A and B be two metagroups, and let $\mathcal{C}$ be a commutative group such that $\mathcal{C}_m(A) \hookrightarrow \mathcal{C}$, $\mathcal{C}_m(B) \hookrightarrow \mathcal{C}$, $\mathcal{C} \hookrightarrow \mathcal{C}(A)$ and $\mathcal{C} \hookrightarrow \mathcal{C}(B)$, where $\mathcal{C}_m(A)$ denotes a minimal subgroup in $\mathcal{C}(A)$ generated by $\{t_A(a,b,c) : a \in A, b \in A, c \in A\}$.*

Using direct products, it is always possible to extend either A or B to get such a case. In particular, either A or B may be a group. Let an equivalence relation Ξ on the Cartesian product $A \times B$ be such that
(2) $(\gamma v, b)\Xi(v, \gamma b)$ *and* $(\gamma v, b)\Xi\gamma(v, b)$ *and* $(\gamma v, b)\Xi(v, b)\gamma$
for every v in A, b in B and γ in $\mathcal{C}$.
(3) *Let $\phi : A \to \mathcal{A}(B)$ be a single-valued mapping, where $\mathcal{A}(B)$ denotes a family of all bijective surjective single-valued mappings of B onto B subjected to the conditions given below. If $a \in A$ and $b \in B$, then b^a is written instead of $\phi(a)b$ for short, where $\phi(a) : B \to B$. Also, let $\eta_\phi : A \times A \times B \to \mathcal{C}$, $\kappa_\phi : A \times B \times B \to \mathcal{C}$ and $\xi_\phi : ((A \times B)/\Xi) \times ((A \times B)/\Xi) \to \mathcal{C}$ be single-valued mappings written as η, κ, and ξ for short, such that*
(4) $(b^u)^v = b^{vu}\eta(v,u,b)$, $e^u = e$, $b^e = b$;
(5) $\eta(v,u,\gamma b) = \eta(v,u,b)$;
(6) $(cb)^u = c^u b^u \kappa(u,c,b)$;
(7) $\kappa(u,\gamma c,b) = \kappa(u,c,\gamma b) = \kappa(u,c,b)$ *and*

$\kappa(u,\gamma,b) = \kappa(u,b,\gamma) = e$;
(8) $\xi((\gamma u,c),(v,b)) = \xi((u,c),(\gamma v,b)) = \xi((u,c),(v,b))$ *and*

$\xi((\gamma,e),(v,b)) = e$ *and* $\xi((u,c),(\gamma,e)) = e$
for every u and v in A, b, c in B, γ in $\mathcal{C}$, where e denotes the neutral element in $\mathcal{C}$ and in A and B.

We put
(9) $(a_1,b_1)(a_2,b_2) = (a_1 a_2, \xi((a_1,b_1),(a_2,b_2))b_1 b_2^{a_1})$
for each a_1, a_2 in A, b_1 and b_2 in B.
The Cartesian product $A \times B$ supplied with such a binary operation (9) is denoted by $A \otimes^{\phi,\eta,\kappa,\xi} B$.

Theorem 9. *Let the conditions of Remark 2 be fulfilled. Then, the Cartesian product $A \times B$ supplied with a binary Operation (9) in Remark 2 is a metagroup.*

Proof. From the conditions of Remark 2, it follows that the binary operation (9) in Remark 2 is single-valued.

Let $I_1 = ((a_1,b_1)(a_2,b_2))(a_3,b_3)$ and $I_2 = (a_1,b_1)((a_2,b_2)(a_3,b_3))$, where a_1, a_2, a_3 belong to A, and b_1, b_2, b_3 belong to B. Then, we infer that
$I_1 = ((a_1a_2)a_3, \xi((a_1,b_1),(a_2,b_2))\xi((a_1a_2, b_1b_2^{a_1}),(a_3,b_3))(b_1b_2^{a_1})b_3^{a_1a_2})$ and
$I_2 = (a_1(a_2a_3), \xi((a_1,b_1),(a_2a_3, b_2b_3^{a_2}))[\xi((a_2,b_2),(a_3,b_3))]^{a_1}$
$b_1(b_2^{a_1}b_3^{a_1a_2})\kappa(a_1,b_2,b_3^{a_2})\eta(a_1,a_2,b_3))$. Therefore,

(1) $I_1 = t_3((a_1,b_1),(a_2,b_2),(a_3,b_3))I_2$ with
(2) $t_3((a_1,b_1),(a_2,b_2),(a_3,b_3)) = t_{3,A}(a_1,a_2,a_3)t_{3,B}(b_1, b_2^{a_1}, b_3^{a_1a_2})$

$\xi((a_1,b_1),(a_2a_3,b_2b_3^{a_2}))[\xi((a_2,b_2),(a_3,b_3))]^{a_1}\kappa(a_1,b_2,b_3^{a_2})\eta(a_1,a_2,b_3)$
$/[\xi((a_1,b_1),(a_2,b_2))\xi((a_1a_2,b_1b_2^{a_1}),(a_3,b_3))]$.

Apparently, $t_{3,A\bigotimes^{\phi,\eta,\kappa,\xi}B}((a_1,b_1),(a_2,b_2),(a_3,b_3)) \in \mathcal{C}$ for each $a_j \in A$, $b_j \in B$, $j \in \{1,2,3\}$, where for shortening of a notation, $t_{3,A\bigotimes^{\phi,\eta,\kappa,\xi}B}$ is denoted by t_3.

If $\gamma \in \mathcal{C}$, then
$\gamma((a_1,b_1)(a_2,b_2)) = (\gamma a_1a_2,\xi((a_1,b_1),(a_2,b_2))b_1b_2^{a_1}) = (a_1a_2,b_1b_2^{a_1})\gamma\xi((a_1,b_1),(a_2,b_2)) = ((a_1,b_1)(a_2,b_2))\gamma$.

Hence, $\gamma \in \mathcal{C}(A\bigotimes^{\phi,\eta,\kappa,\xi}B)$. Consequently, $\mathcal{C} \subseteq \mathcal{C}(A\bigotimes^{\phi,\eta,\kappa,\xi}B)$.

Next, we consider the following equation:

(3) $(a_1,b_1)(a,b) = (e,e)$, where $a \in A$, $b \in B$.

From (2) of Definition 1 in Section 2 and (9) in Remark 2, we deduce that

(4) $a_1 = e/a$.
Consequently, $\xi((e/a,b_1),(a,b))b_1b^{(e/a)} = e$, and hence

(5) $b_1 = e/[\xi((e/a,b^{(e/a)}),(a,b))b^{(e/a)}]$.

Thus, $a_1 \in A$ and $b_1 \in B$ given by (4) and (5) provide a unique solution of (3).

Similarly, from the following equation

(6) $(a,b)(a_2,b_2) = (e,e)$, where $a \in A$, $b \in B$ we infer that

(7) $a_2 = a\setminus e$.
Consequently, $\xi((a,b),(a\setminus e,b_2))bb_2^a = e$, and hence, $b_2^a = [\xi((a,b),(a\setminus e,b_2))b]\setminus e$. On the other hand, $(b_2^a)^{e/a} = \eta(e/a,a,b_2)b_2$ Consequently,

(8) $b_2 = (b\setminus e)^{e/a}/\{[(\xi((a,b),(a\setminus e,(b\setminus e)^{e/a}))]^{e/a}\eta(e/a,a,(b\setminus e)^{e/a})\}$.

Thus, Formulas (7) and (8) provide a unique solution to (6).

Next, we put $(a_1,b_1) = (e,e)/(a,b)$ and $(a_2,b_2) = (a,b)\setminus(e,e)$ and

(9) $(a,b)\setminus(c,d) = ((a,b)\setminus(e,e))(c,d)$
$t_3((e,e)/(a,b),(a,b),((a,b)\setminus(e,e))(c,d))/t_3((e,e)/(a,b),(a,b),(a,b)\setminus(e,e))$;

(10) $(c,d)/(a,b) = (c,d)((e,e)/(a,b))$
$t_3((e,e)/(a,b),(a,b),(a,b)\setminus(e,e))/t_3((c,d)(e/(a,b)),(a,b),(a,b)\setminus(e,e))$

and $e_G = (e,e)$, where $G = A\bigotimes^{\phi,\eta,\kappa,\xi}B$. Note that (3) of Definition 1 in Section 2 follows on from (8) and (9) of Remark 2. Therefore, Properties (1)–(3) and (9) of Definition 1 in Section 2 are fulfilled for $A\bigotimes^{\phi,\eta,\kappa,\xi}B$. □

Definition 5. *The metagroup $A\bigotimes^{\phi,\eta,\kappa,\xi}B$ provided by Theorem 9 is called a smashed product of metagroups A and B with smashing factors ϕ, η, κ, and ξ.*

Remark 3. *From Theorems 8 and 9, it follows that by taking the nontrivial mappings η, κ, and ξ and starting from groups with nontrivial $C(G_j)$ or $C(A)$, it is possible to construct new metagroups with nontrivial $C(G)$, and ranges $t_{3,G}(G,G,G)$ of $t_{3,G}$ may be infinite. Indeed, using extensions of groups (or metagroups) by semidirect or direct products, it is possible to take initial groups (or metagroups) A and B such that quotient groups A_1 of A by $C(A)$ and B_1 of B by $C(B)$ are infinite. Therefore, their automorphism groups $Aut(A_1)$ and $Aut(B_1)$ are infinite, because they contain all inner automorphisms.*

With suitable smashing factors ϕ, η, κ, and ξ and with nontrivial metagroups or groups A and B it is easy to get examples of metagroups in which $e/a \neq a\setminus e$ for an infinite family of elements a in $A\bigotimes^{\phi,\eta,\kappa,\xi}B$ using Formulas (1) in Lemma 2 in Section 2 and (2) in Theorem 9. Evidently smashed products (see Remark 2 and Theorem 9) are nonassociative generalizations of semidirect products.

Smashed twisted products and smashed twisted wreath products of metagroups or groups are described in [34]. They also provide tools for a construction of a wide class of metagroups and nonassociative algebras with metagroup relations.

Conclusions 1. The results of this article can be used for further studies of cohomology theory of nonassociative algebras and noncommutative manifolds with metagroup relations. It is interesting to mention possible applications in mathematical coding theory, analysis of information flows, and their technical realizations [35–38], because codes are frequently based on binary systems and algebras. Indeed, metagroup relations are weaker than relations in groups. Therefore, a code complexity can increase by using nonassociative algebras with metagroup relations in comparison with group algebras or Lie algebras.

Besides the applications of cohomologies outlined in the introduction, they also can be used in mathematical physics and quantum field theory. This also can be applied to cohomologies of PDEs and solutions of PDEs with boundary conditions, which can have practical importance [27,39]. It will be interesting to investigate cohomologies of nonassociative algebras related with a class of directed ringoids [40,41], because the latter have applications to non locally compact groups.

Funding: This research received no external funding.

Conflicts of Interest: The author declares no conflict of interest.

References

1. Cartan, H.; Eilenberg, S. *Homological Algebra*; Princeton University Press: Princeton, NJ, USA, 1956.
2. Goto, M.; Grosshans, F.D. *Semisimple Lie algebras*; Marcel Dekker, Inc.: New York, NY, USA, 1978.
3. Shang, Y. A Lie algebra approach to susceptible-infected-susceptible epidemics. *Electron. J. Differ. Equat.* **2012**, *2012*, 1–7.
4. Shang, Y. Analytic solution for an in-host viral invection model with time-inhomogeneous rates. *Acta Phys. Pol.* **2015**, *46*, 1567–1577. [CrossRef]
5. Shang, Y. Lie algebraic discussion for affinity based information diffusion in social networks. *Open Phys.* **2017**, *15*, 705–711. [CrossRef]
6. Allcock, D. Reflection groups and octave hyperbolic plane. *J. Algebra* **1998**, *213*, 467–498. [CrossRef]
7. Baez, J.C. The octonions. *Bull. Am. Math. Soc.* **2002**, *39*, 145–205. [CrossRef]
8. Dickson, L.E. *The Collected Mathematical Papers*; Chelsea Publishing Co.: New York, NY, USA, 1975; Volumes 1–5.
9. Kantor, I.L.; Solodovnikov, A.S. *Hypercomplex Numbers*; Springer: Berlin, Germany, 1989.
10. Schafer, R.D. *An Introduction to Nonassociative Algebras*; Academic Press: New York, NY, USA, 1966.
11. Serôdio, R. On octonionic polynomials. *Adv. Appl. Clifford Algebras* **2007**, *17*, 245–258.
12. Frenod, E.; Ludkowski, S.V. Integral operator approach over octonions to solution of nonlinear PDE. *Far East J. Math. Sci. (FJMS)* **2018**, *103*, 831–876. [CrossRef]
13. Girard, P.R. *Quaternions, Clifford Algebras and Relativistic Physics*; Birkhäuser: Basel, Switzerland, 2007.
14. Gürlebeck, K.; Sprössig, W. *Quaternionic and Clifford Calculus for Physicists and Engineers*; John Wiley and Sons: Chichester, NH, USA, 1997.
15. Gürsey, F.; Tze, C.-H. *On the Role of Division, Jordan and Related Algebras in Particle Physics*; World Scientific Publishing Co.: Singapore, 1996.
16. Krausshar, R.S. *Generalized Analytic Automorphic Forms in Hypercomplex Spaces*; Birkhäuser: Basel, Switzerland, 2004.
17. Ludkovsky, S.V. Wrap groups of connected fiber bundles: Their structure and cohomologies. *Lie Groups: New Research*; Canterra, A.B., Ed.; Nova Science Publishers, Inc.: New York, NY, USA, 2009; pp. 53–128.
18. Ludkowski, S.V. Decompositions of PDE over Cayley-Dickson algebras. *Rend. Istit. Mat. Univ.* **2014**, *46*, 1–23.
19. Ludkowski, S.V. Integration of vector Sobolev type PDE over octonions. *Complex Var. Elliptic Equat.* **2016**, *61*, 1014–1035.
20. Ludkowski, S.V. Manifolds over Cayley-Dickson algebras and their immersions. *Rend. Istit. Mat. Univ.* **2013**, *45*, 11–22.

21. Ludkovsky, S.V. Normal families of functions and groups of pseudoconformal diffeomorphisms of quaternion and octonion variables. *J. Math. Sci. N. Y.* **2008**, *150*, 2224–2287. [CrossRef]
22. Ludkovsky, S.V. Functions of several Cayley-Dickson variables and manifolds over them. *J. Math. Sci. N. Y.* **2007**, *141*, 1299–1330. [CrossRef]
23. Ludkovsky, S.V.; Sprössig, W. Ordered representations of normal and super-differential operators in quaternion and octonion Hilbert spaces. *Adv. Appl. Clifford Alg.* **2010**, *20*, 321–342.
24. Ludkovsky, S.V.; Sprössig, W. Spectral theory of super-differential operators of quaternion and octonion variables. *Adv. Appl. Clifford Alg.* **2011**, *21*, 165–191.
25. Ludkovsky, S.V. Integration of vector hydrodynamical partial differential equations over octonions. *Complex Var. Elliptic Equat.* **2013**, *58*, 579–609. [CrossRef]
26. Ludkowski, S.V. Automorphisms and derivations of nonassociative C^* algebras. *Linear Multilinear Algebra* **2018**, 1–8. [CrossRef]
27. Pommaret, J.F. *Systems of Partial Differential Equations and Lie Pseudogroups*; Gordon and Breach Science Publishers: New York, NY, USA, 1978.
28. Bourbaki, N. Algèbre. In *Algèbre Homologique*; Springer: Berlin, Germany, 2007.
29. Bredon, G.E. *Sheaf Theory*; McGarw-Hill: New York, NY, USA, 2012.
30. Hochschild, G. On the cohomology theory for associative algebras. *Ann. Math.* **1946**, *47*, 568–579. [CrossRef]
31. Chapoton, F.; Livernet, M. Pre-Lie algebras and rooted trees operad. *Int. Math. Res. Not.* **2001**, *8*, 395–408. [CrossRef]
32. Dzhumadil'daev, A.; Zusmanovich, P. The alternative operad is not Koszul. *Exper. Math.* **2011**, *20*, 138–144. [CrossRef]
33. Remm, E.; Goze, M. A class of nonassociative algebras including flexible and alternative algebras, operads and deformations. *J. Gener. Lie Theory Appl.* **2015**, *9*, 1–6.
34. Ludkowski, S.V. Metagroups and their smashed twisted wreath products. *arXiv* **2018**, arXiv:1809.02801.
35. Blahut, R.E. *Algebraic Codes for Data Transmission*; Cambridge University Press: Cambridge, UK, 2003.
36. Magomedov, S.G. Assessment of the impact of confounding factors in the performance information security. *Rus. Technol. J.* **2017**, *5*, 47–56.
37. Sigov, A.S.; Andrianova, E.G.; Zhukov, D.O.; Zykov, S.V.; Tarasov, I.E. Quantum informatics: Overview of the main achievements. *Rus. Technol. J.* **2019**, *7*, 5–37. [CrossRef]
38. Shum, K.P.; Ren, X.; Wang, Y. Semigroups on semilattice and the constructions of generalized cryptogroups. *Southeast Asian Bull. Math.* **2014**, *38*, 719–730.
39. Zaikin, B.A.; Bogadarov, A.Y..; Kotov, A.F.; Poponov, P.V. Evaluation of coordinates of air target in a two-position range measurement radar. *Rus. Technol. J.* **2016**, *4*, 65–72.
40. Ludkovsky, S.V. On transfinite construction of a class of directed topological ringoids. *JP J. Algebra Number Theory Appl.* **2015**, *37*, 185–208. [CrossRef]
41. Ludkowski, S.V. Skew continuous morphisms of ordered lattice ringoids. *Mathematics* **2016**, *4*, 17. [CrossRef]

axioms

MDPI

Article

Separability of Nonassociative Algebras with Metagroup Relations

Sergey V. Ludkowski

Department of Applied Mathematics, MIREA—Russian Technological University, av. Vernadsky 78, 119454 Moscow, Russia; sludkowski@mail.ru

Received: 15 November 2019; Accepted: 10 December 2019; Published: 12 December 2019

Abstract: This article is devoted to a class of nonassociative algebras with metagroup relations. This class includes, in particular, generalized Cayley–Dickson algebras. The separability of the nonassociative algebras with metagroup relations is investigated. For this purpose the cohomology theory is utilized. Conditions are found under which such algebras are separable. Algebras satisfying these conditions are described.

Keywords: algebra; nonassociative; separable; ideal; cohomology

Mathematics Subject Classification 2010: 17A30; 17A60; 17A70; 14F43; 18G60

---

1. Introduction

Associative separable algebras play an important role and have found many-sided application (see, for example, [1–9] and references therein). Studies of their structure are based on cohomology theory. On the other hand, cohomology theory of associative algebras was investigated by Hochschild and other authors [10–13], but it is not applicable to nonassociative algebras. Cohomology theory of group algebras is an important and great part of algebraic topology.

It is worth mentioning that nonassociative algebras with some identities in them found many-sided applications in physics, noncommutative geometry, quantum field theory, partial differential equations (PDEs) and other sciences (see [14–25] and references therein).

An extensive area of investigations of PDEs intersects with cohomologies and deformed cohomologies [13]. Therefore, it is important to develop this area over octonions, generalized Cayley–Dickson algebras and more general metagroup algebras (see also Appendix A). Some results in this area are presented in [26]. The structure of metagroups, their construction and examples, and smashed and twisted wreath products were studied and described in [26–28]. In particular, a class of metagroup algebras contains a family of generalized Cayley–Dickson algebras and nonassociative analogs of C^* algebras.

For comparison it is worth noting that there are algebras with relations T induced by Jordan-type or Lie-type homomorphisms in the sense of [29]. Their unified approach (UJLA) was studied in [22]. In those works the unital universal envelope $U_R(A)$ of a nonassociative algebra A with relations T was considered, where R denotes an associative commutative unital ring. The algebra $U_R(A)$ is associative and may be noncommutative. This theory is applicable to Lie algebras, alternative algebras, Jordan algebras and UJLA fitting to algebras with relations T.

However, this technique is not applicable to the metagroup algebras studied in this article. Indeed, there are several obstacles. The algebra $U_R(A)$ is associative and with it a lot of information about the metagroup algebras is lost. A derivation functor cannot serve as a starting point for a construction of a cohomology theory for the metagroup algebras. Moreover relations in metagroup algebras are external to them and do not fit to the nonassociative algebras with relations T considered in [22,29].

This article is devoted to a separability of nonassociative algebras with metagroup relations. Conditions are found under which they are separable. Algebras satisfying these conditions are described in Theorems 1–3.

All main results of this paper are obtained for the first time.

2. Separable Nonassociative Algebras

Nonassociative metagroups, their centers, metagroup algebras and modules over them were defined in [26–28] (see also Appendix A). To avoid misunderstandings we also give specific necessary definitions and notations.

Definition 1. *Let Ψ be a (proper or improper) subgroup in the center $\mathcal{C}(G)$ of a metagroup G, let 1 denote a unit in $\mathcal{T}$, e be a unit in G and let*

A be a nonassociative metagroup algebra over a commutative associative unital ring $\mathcal{T}$ such that

$$\Psi 1 \subseteq (G1) \cap \mathcal{T}e, \tag{1}$$

where $(G1) \cup \mathcal{T}e \subset A$, $A = \mathcal{T}[G]$ denotes a metagroup algebra.

A G-graded A-module P (also see Definition 3 in [26]) is called projective if it is isomorphic with a direct summand of a free G-graded A-module. The metagroup algebra A is called separable if it is a projective G-graded A^e-module.

One puts $\mu(z) = 1_A z$ for each $z \in A^e$, where A is considered as the G-graded right A^e-module.

Proposition 1. *Suppose that A is a nonassociative algebra satisfying condition (1). Then the following conditions are equivalent:*

$$A \textit{ is separable} \tag{2}$$

the exact sequence

$$0 \to Ker\,\mu \to A^e \underset{\mu}{\to} A \to 0 \textit{ splits} \tag{3}$$

$$\textit{an element } b \in A^e \textit{ exists such that } \mu(b) = 1_A \textit{ and } xb = bx \textit{ and } b(xy) = (bx)y \textit{ and } (xb)y = x(by) \textit{ and } (xy)b = x(yb) \textit{ for all } x \textit{ and } y \textit{ in } A, \tag{4}$$

where A^e is considered as the G-graded two-sided A-module.

Proof. The implication (2) $\Rightarrow$ (3) is evident.

(3) $\Rightarrow$ (4). If the exact sequence (3) splits, then A^e as the A^e-module is isomorphic with $A \oplus ker(\mu)$, where $\oplus$ denotes a direct sum. Therefore, A is separable. The sequence (3) splits if and only if there exists $p \in Hom_{A^e}(A, A^e)$ such that $\mu p = id_A$. With this homomorphism p put $b = p(1_A)$. Then $(xb)y = (xp(1_A))y = p(x1_A)y = p(x(1_Ay)) = p((xy)1_A) = (xy)b$, hence $\mu(b) = \mu p(1_A) = 1_A$ and $xb = xp(1_A) = p(x1_A) = p(1_Ax) = p(1_A)x = bx$. Thus (4) is valid.

(4) $\Rightarrow$ (2). Suppose that condition (4) is fulfilled. Then a mapping $p : A \to A^e$ exists such that $p(x) = bx$. The element b has the decomposition $b = \sum_j b_j g_j$ with $g_j = g_{j,1} \otimes g_{j,2}$, where $g_{j,1} \in G$ and $g_{j,2} \in G^{op}$ and $b_j \in \mathcal{T}$ for each j. Therefore, using condition (4) above and conditions (1)–(3) in Definition 3 in [26] we infer that

$$p(xy) = \sum_j \sum_k \sum_l b_j g_j((c_k x_k)(d_l y_l)) = \sum_j b_j (g_j x) y = (bx)y = p(x)y$$

and

$$p(yx) = (by)x = (yb)x = y(bx) = yp(x)$$

for each x and $y \in A$, where $x = \sum_k c_k x_k$ and $y = \sum_l d_l y_l$ with x_k and y_l in G, c_k and d_l in $\mathcal{T}$ for each k and l. Thus $p \in Hom_{A^e}(A, A^e)$. Moreover, $\mu(p(x)) = \mu(bx) = \mu(b)x = 1_A x = x$ for each $x \in A$, consequently, the exact sequence (3) splits. □

Definition 2. *An element $b \in A^e$ fulfilling condition (4) in Proposition 1 is called a separating idempotent of an algebra A.*

Lemma 1. *Let A be a nonassociative algebra satisfying condition (1). Let also M be a two-sided A-module.*

If $p \in Hom_{A^e}(ker(\mu), M)$ and $\kappa : A \to A^e$ with $\kappa(x) = x \otimes 1 - 1 \otimes x$ for each $x \in A$, then $p\kappa$ is a derivation of A with values in M. (5)

$$\text{A mapping } \chi : p \mapsto p\kappa \text{ is an isomorphism of } Hom_{A^e}(ker(\mu), M) \text{ onto } Z^1_{\mathcal{T}}(A, M). \tag{6}$$

$$\chi^{-1}(B^1_{\mathcal{T}}(A, M)) = \{\psi|_{\ker(\mu)} : \psi \in Hom_{A^e}(A^e, M)\}. \tag{7}$$

Proof. (5). Since $\mu\kappa = 0$, then $Im(\kappa) \subseteq ker(\mu)$. By virtue of Theorem 1 in [26] $\mu\kappa$ is the derivation having also properties (6) and (7). □

Theorem 1. *Suppose that A is a nontrivial nonassociative algebra satisfying condition (1). Then $H^1_{\mathcal{T}}(A, M) = 0$ for each two-sided A-module M if and only if A is a separable $\mathcal{T}$-algebra.*

Proof. In view of Proposition 1 the algebra A is separable if and only if the exact sequence (3) splits. That is, a homomorphism h exists $h \in Hom_{A^e}(A, ker(\mu))$ such that its restriction $h|_{ker(\mu)}$ is the identity mapping. Therefore, if $H^1_{\mathcal{T}}(A, ker(\mu)) = 0$, then the algebra A is separable due to Lemma 1.

Vice versa if a homomorphism $h \in Hom_{A^e}(A^e, ker(\mu))$ exists with $h|_{ker(\mu)} = id$, then each $p \in Hom_{A^e}(ker(\mu), M)$ has the form $f|_{ker(\mu)}$ with $f = ph \in Hom_{A^e}(A^e, M)$. By virtue of Lemma 1 $Z^1_{\mathcal{T}}(A, M) = B^1_{\mathcal{T}}(A, M)$ for each two-sided A-module M. □

Theorem 2. *Let a noncommutative algebra A fulfill condition (1) and*

$$Dim(A/J(A)) < 1 \tag{8}$$

$$\text{and } A/J(A) \text{ is projective as the } \mathcal{T}\text{-module} \tag{9}$$

$$\text{and } J(A)^k = 0 \text{ for some } k \geq 1, \tag{10}$$

where $J(A)$ denotes the radical of A.

Then a subalgebra D in A exists such that $A = D \oplus J(A)$ as $\mathcal{T}$-modules and $A/J(A)$ is isomorphic with D as the algebra.

Proof. For $k = 1$ we get $A = D$.

For $k = 2$ a natural projection $\pi : A \to A/J$ exists, where $J = J(A) = rad(A_A)$, since $J^2 = 0$. The algebra A is G-graded and $\mathcal{T} \subseteq Z(A)$, hence $rad((A_e)_{A_e}) \subseteq (rad(A_A))_e$, where e is the unit element of G. In view of conditions (1)–(3) in Definition 3 in [26] J is the two-sided ideal in A and $J^m_r = J^m_l$ for each positive integer m, where $J^1_l = J$, $J^1_r = J$, $J^{m+1}_l = JJ^m_l$ and $J^{m+1}_r = J^m_r J$. Condition (4) in Definition 1 in [27] and conditions (1)–(3) in Definition 3 in [26] imply that A/J is also G-graded, since $\mathcal{T} \subset Z(A)$.

By condition (9) the $\mathcal{T}$-module A/J is projective, consequently, an exact splitting sequence of $\mathcal{T}$-modules exists

$$0 \to J \to A \to A/J \to 0. \tag{11}$$

Thus a homomorphism $\kappa : A/J \to A$ of $\mathcal{T}$-modules exists such that $\pi\kappa = id$ on A/J. For any two elements x and y in A/J we put

$$\Phi(x,y) = \kappa(xy) - \kappa(x)\kappa(y) \tag{12}$$

Therefore, we infer that

$$\pi\Phi(x,y) = \pi\kappa(xy) - \pi(\kappa(x)\kappa(y)) = xy - xy = 0 \tag{13}$$

since π is the algebra homomorphism and $\pi\kappa = id$. Thus $\Phi(x,y) \in ker(\pi) = J$. One has by the definition that

$$Dim(A/J) = \sup\{n : \exists \text{ two-sided } A/J\text{-module } M \ \ H^n_{\mathcal{T}}(A/J, M) \neq 0\} \tag{14}$$

Then put $ux = u\kappa(x)$ and $xu = \kappa(x)u$ to be the right and left actions of A/J on J. Since κ is the homomorphism of $\mathcal{T}$-modules and $\mathcal{T} \subseteq Z(A)$, then for each pure states x, y and u we infer:

$$(xy)u - \mathrm{t}_3 x(yu) = \kappa(xy)u - (\kappa(x)\kappa(y))u = \Phi(x,y)u \in J^2 = 0 \tag{15}$$

where $\mathrm{t}_3 = \mathrm{t}_3(x,y,u)$. Then we deduce that

$$u(xy) - \mathrm{t}_3^{-1}(ux)y = u\kappa(xy) - u(\kappa(x)\kappa(y)) = u\Phi(x,y) \in J^2 = 0 \tag{16}$$

where $\mathrm{t}_3 = \mathrm{t}_3(u,x,y)$. Thus J has the structure of the two-sided A/J-module.

Evidently, Φ is $\mathcal{T}$-bilinear. Then for every pure states x, y and z in A/J:

$$\begin{aligned}(\delta^2\Phi)(x,y,z) = \mathrm{t}_3 x(\kappa(yz) - \kappa(y)\kappa(z)) - (\kappa((xy)z) - \kappa(xy)\kappa(z)) + \\ \mathrm{t}_3(\kappa(x(yz)) - \kappa(x)\kappa(yz)) - (\kappa(xy) - \kappa(x)\kappa(y))z \\ = \mathrm{t}_3\kappa(x)\kappa(yz) - \mathrm{t}_3\kappa(x)(\kappa(y)\kappa(z)) - \kappa((xy)z) + \kappa(xy)\kappa(z) \\ +\mathrm{t}_3\kappa(x(yz)) - \mathrm{t}_3\kappa(x)\kappa(yz) - \kappa(xy)\kappa(z) + (\kappa(x)\kappa(y))\kappa(z) = 0\end{aligned} \tag{17}$$

consequently, $\Phi \in B^2_{\mathcal{T}}(A/J, J)$, where $\mathrm{t}_3 = \mathrm{t}_3(x,y,z)$. Thus by the $\mathcal{T}$-linearity a homomorphism h in $Hom_{\mathcal{T}}(A/J, J)$ exists possessing the property

$$\Phi(x,y) = xh(y) - h(xy) - h(x)y \tag{18}$$

for each x and y in A/J.

Let now $p = \kappa + h \in Hom_{\mathcal{T}}(A/J, J)$, consequently, $\pi p = \pi\kappa = id|_{A/J}$, since $\pi(J) = 0$. This implies that $p(xy) - p(x)p(y) = 0$ for each x and y in A/J, since $\kappa(xy) - \kappa(x)\kappa(y) = \Phi(x,y) = xh(y) - h(xy) + h(x)y$ and $h(x)h(y) \in J^2 = 0$. Since $p(1_{A/J}) - 1_A \in J$, then $(p(1) - 1)^2 = 1 - p(1)$. Therefore, p is the algebra homomorphism. This implies that $D = Im(p)$ is the subalgebra in A such that $A = D \oplus J$.

Let now $k > 2$ and this theorem is proven for $1, \ldots, k-1$. Put $A_1 = A/J^2$, then J/J^2 is the two-sided ideal in A_1 and $A_1/(J/J^2)$ is isomorphic with A/J, also $(J/J^2)^2 = 0$. Thus $J(A_1) = J/J^2$ and A_1 satisfies conditions (8)–(10) of this theorem and is G-graded, since A and J are G-graded and $\mathcal{T} \subset Z(G)$ due to condition (4) in Definition 1 in [27] and conditions (1)–(3) in Definition 3 in [26].

From the proof for $k = 2$ we get that a subalgebra D_1 in A_1 exists such that $A_1 = D_1 \oplus J/J^2$. Consider a subalgebra E in D such that $E \cap J = J^2$ and $D_1 = E/J^2$. Then E/J is isomorphic with $E/(E \cap J) \approx (E + J)/J = A/J$. Moreover, $(J^2)^{k-1} = J^{k+k-2} \subseteq J^k = 0$, hence $J(E) = J^2$. Thus the algebra E fulfills conditions (8)–(10) of this theorem and is G-graded and $J(E)^{k-1} = 0$.

By the induction supposition a subalgebra F in E exists such that $E = F \oplus J^2$; consequently, $F + J = E + J = A$ and $F \cap J = F \cap E \cap J = F \cap J^2 = 0$. Thus $A = F \oplus J$. □

Theorem 3. *Suppose that conditions of Theorem 2 are satisfied and condition (8) takes the form* $Dim(A/J(A)) = 0$. *Then for any two G-graded subalgebras B and C in A such that* $A = B \oplus J(A)$ *and* $A = C \oplus J(A)$ *an element* $v \in J(A)$ *exists for which* $(1-v)C = B(1-v)$ *such that* $(1-v)$ *has a right inverse and a left inverse.*

Proof. Let $q : A \to B$ and $r : A \to C$ be the canonical projections induced by the decompositions $A = B \oplus J$ and $A = C \oplus J$, where $J = J(A)$. Then $p\pi = q$ and $s\pi = r$, where $\pi : A \to A/J$ is the quotient homomorphism, $p : A/J \to A$ and $s : A/J \to C$ are homomorphisms as in the proof of Theorem 2, since q and r are homomorphisms of algebras. We put $w(x) = p(x) - s(x)$ for each $x \in A/J$, $w : A/J \to J$. Then we deduce that

$$\pi(w\pi) = \pi(p\pi) - \pi(s\pi) = \pi q - \pi r = \pi(id_A - r) - \pi(id_A - q) = 0, \tag{19}$$

since $Im(id_A - q) = Im(id_A - r) = J = ker(\pi)$. Therefore, $Im(w) = Im(w\pi) \subseteq J$, hence $w \in Hom_{\mathcal{T}}(A/J, J)$. Then we infer that

$$\begin{gathered} w(xy) = p(xy) - s(xy) = \\ p(x)(p(y) - s(y)) + (p(x) - s(x))s(y) = xw(y) + w(x)y \end{gathered} \tag{20}$$

consequently, w is the derivation of the algebra A/J with values in the two-sided A-module A/J (see also the proof of Theorem 2). Since $Dim(A/J) = 0$, then w is the inner derivation by Theorem 1 in [26]. Thus an element $v \in J$ exists for which $w(x) = xv - vx$ for each $x \in A/J$. This implies that $p(x)(1-v) = (1-v)s(x)$ for each $x \in A/J$. The element $(1-v)$ has a right inverse and a left inverse, since $J^k = 0$ implies $v_l^k = 0$ and $v_r^k = 0$, where $v_l^1 = v$, $v_r^1 = v$, $v_l^{m+1} = vv_l^m$ and $v_r^{m+1} = v_r^m v$ for each positive integer m. Therefore,

$$B(1-v) = p(A/J)(1-v) = (1-v)s(A/J) = (1-v)C \tag{21}$$

□

Remark 1. *Definition 1 is natural. For example, if* $\mathcal{J}$ *is a commutative associative unital ring and S is a subgroup in* $C(G)$, *then* $\mathcal{T}_1 := \mathcal{J}[S]$ *is a commutative associative unital ring such that* $S1 \subseteq G1 \cap \mathcal{T}_1 e$.

3. Conclusions

The results of this article can be used for further studies of nonassociative algebras, their structure, cohomologies, algebraic geometry, PDEs, their applications in the sciences, etc. They also can serve for investigations of extensions of nonassociative algebras, decompositions of algebras and modules, and their morphisms. In particular, they can be applied to cohomologies of PDEs and solutions of PDEs with boundary conditions which can have a practical importance [13,30].

Other applications are in mathematical coding theory, information flows analysis and their technological implementations [31–34]. Indeed, frequently codes are based on binary systems and algebras. On the other hand, metagroup relations are weaker than relations in groups. This means that a code complexity can increase by using nonassociative algebras with metagroup relations in comparison with group algebras or Lie algebras.

Besides applications of cohomologies outlined in the introduction they also can be used in mathematical physics and quantum field theory [15].

Funding: This research received no external funding.

Conflicts of Interest: The author declares no conflict of interest.

Appendix A. Metagroups

Let G be a set with a single-valued binary operation (multiplication) $G^2 \ni (a,b) \mapsto ab \in G$ defined on G satisfying the conditions:

$$\text{for each } a \text{ and } b \text{ in } G \text{ there is a unique } x \in G \text{ with } ax = b \tag{A1}$$

$$\text{and a unique } y \in G \text{ exists satisfying } ya = b \tag{A2}$$

which are denoted by

$x = a \setminus b = Div_l(a,b)$ and $y = b/a = Div_r(a,b)$ correspondingly,

there exists a neutral (i.e., unit) element $e_G = e \in G$:

$$eg = ge = g \text{ for each } g \in G. \tag{A3}$$

The set of all elements $h \in G$ commuting and associating with G:

$$Com(G) := \{a \in G : \forall b \in G,\ ab = ba\}, \tag{A4}$$

$$N_l(G) := \{a \in G : \forall b \in G, \forall c \in G,\ (ab)c = a(bc)\}, \tag{A5}$$

$$N_m(G) := \{a \in G : \forall b \in G, \forall c \in G,\ (ba)c = b(ac)\}, \tag{A6}$$

$$N_r(G) := \{a \in G : \forall b \in G, \forall c \in G,\ (bc)a = b(ca)\}, \tag{A7}$$

$$N(G) := N_l(G) \cap N_m(G) \cap N_r(G); \tag{A8}$$

$$\mathcal{C}(G) := Com(G) \cap N(G) \tag{A9}$$

is called the center $\mathcal{C}(G)$ of G.

We call G a metagroup if a set G possesses a single-valued binary operation and satisfies conditions (A1)–(A3) and

$$(ab)c = t_3(a,b,c)a(bc) \tag{A10}$$

for each a, b and c in G, where $t_3(a,b,c) \in \mathbf{\Psi}$, $\mathbf{\Psi} \subset \mathcal{C}(G)$; where t_3 shortens a notation $t_{3,G}$, where $\mathbf{\Psi}$ denotes a (proper or improper) subgroup of $\mathcal{C}(G)$.

Then G will be called a central metagroup if in addition to (A10) it satisfies the condition:

$$ab = t_2(a,b)ba \tag{A11}$$

for each a and b in G, where $t_2(a,b) \in \mathbf{\Psi}$.

From conditions (1)–(3) in Definition 3 in [26] it follows that for each a and b in the metagroup algebra $A = \mathcal{T}[G]$ and x in a (smashly G-graded) two-sided A-module M there may exist a $\mathcal{T}$-homomorphism $P_1(a,b,x) : M' \to M''$ of right $\mathcal{T}$-modules $M' := a(bM)$ and $M'' := (ab)M$ such that $[P_1(a,b,x)]a(bx) = (ab)x$ for chosen a, b and x. Similar homomorphisms $P_2(a,x,b)$ and $P_3(x,a,b)$ may exist on $a(Mb)$ and $M(ab)$, respectively. Generally these homomorphisms $P_1(a,b,x)$, $P_2(a,x,b)$ and $P_3(x,a,b)$ depend nontrivially on all variables a, b and x (see also Remark 1 in [26]). So they cannot be realized by identities of Jordan-type or Lie-type or UJLA-type (see also the introduction).

References

1. Bourbaki, N. *Algèbre*; Springer: Berlin, Germany, 2007; Chapter 1–3.
2. Bourbaki, N. Algèbre homologique. In *Algèbre*; Springer: Berlin, Germany, 2007; Chapter 10.
3. Florence, M. On higher trace forms of separable algebras. *Arch. Math.* **2011**, *97*, 247–249. [CrossRef]
4. Georgantas, G.T. Derivations in central separable algebras. *Glasgow Math. J.* **1978**, *19*, 75–77. [CrossRef]
5. Mazur, M.; Petrenko, B.V. Separable algebras over infinite fields are 2-generated and finitely presented. *Arch. Math.* **2009**, *93*, 521–529. [CrossRef]
6. Montgomery, S.; Smith, M.K. Algebras with a separable subalgebra whose centralizer satisfies a polynomial identity. *Commun. Algebra* **1975**, *3*, 151–168. [CrossRef]
7. Van Oystaeyen, F. Separable algebras. In *Handbook of Algebra*; Hazewinkel, M., Ed.; Elsevier: Amsterdam, The Netherlands, 2000; Volume 2, pp. 463–505.
8. Pierce, R.S. *Associative Algebras*; Springer: New York, NY, USA, 1982.
9. Rumynin, D.A. Cohomomorphisms of separable algebras. *Algebra Log.* **1994**, *33*, 233–237. [CrossRef]
10. Bredon, G.E. *Sheaf Theory*; McGraw-Hill: New York, NY, USA, 2012.
11. Cartan, H.; Eilenberg, S. *Homological Algebra*; Princeton University Press: Princeton, NJ, USA, 1956.
12. Hochschild, G. On the cohomology theory for associative algebras. *Ann. Mathem.* **1946**, *47*, 568–579. [CrossRef]
13. Pommaret, J.F. *Systems of Partial Differential Equations and Lie Pseudogroups*; Gordon and Breach Science Publishers: New York, NY, USA, 1978.
14. Dickson, L.E. *The Collected Mathematical Papers*; Chelsea Publishing Co.: New York, NY, USA, 1975; Volume 1–5.
15. Gürsey, F.; Tze, C.-H. *On the Role of Division, Jordan and Related Algebras in Particle Physics*; World Scientific Publication Co.: Singapore, 1996.
16. Kantor, I.L.; Solodovnikov, A.S. *Hypercomplex Numbers*; Springer: Berlin, Germany, 1989.
17. Krausshar, R.S. *Generalized Analytic Automorphic Forms in Hypercomplex Spaces*; Birkhäuser: Basel, Switzerlnad, 2004.
18. Ludkowski, S.V. Integration of vector Sobolev type PDE over octonions. *Complex Var. Elliptic Equat.* **2016**, *61*, 1014–1035.
19. Ludkovsky, S.V. Normal families of functions and groups of pseudoconformal diffeomorphisms of quaternion and octonion variables. *J. Math. Sci. N. Y.* **2008**, *150*, 2224–2287. [CrossRef]
20. Ludkovsky, S.V.; Sprössig, W. Ordered representations of normal and super-differential operators in quaternion and octonion Hilbert spaces. *Adv. Appl. Clifford Alg.* **2010**, *20*, 321–342.
21. Ludkovsky, S.V.; Sprössig, W. Spectral theory of super-differential operators of quaternion and octonion variables. *Adv. Appl. Clifford Alg.* **2011**, *21*, 165–191.
22. Nichita, F.F. Unification theories: New results and examples. *Axioms* **2019**, *8*, 60. [CrossRef]
23. Schafer, R.D. *An Introduction to Nonassociative Algebras*; Academic Press: New York, NY, USA, 1966.
24. Shang, Y. Lie algebraic discussion for affinity based information diffusion in social networks. *Open Phys.* **2017**, *15*, 705–711. [CrossRef]
25. Shang, Y. A Lie algebra approach to susceptible-infected-susceptible epidemics. *Electr. J. Differ. Equat.* **2012**, *233*, 1–7.
26. Ludkowski, S.V. Cohomology theory of nonassociative algebras with metagroup relations. *Axioms* **2019**, *8*, 78 [CrossRef]
27. Ludkowski, S.V. Automorphisms and derivations of nonassociative C^* algebras. *Linear Multilinear Algebra* **2019**, *67*, 1531–1538 [CrossRef]
28. Ludkowski, S.V. Smashed and twisted wreath products of metagroups. *Axioms* **2019**, *8*, 127. [CrossRef]
29. Jacobson, N. *Structure and Representations of Jordan Algebras*; Colloquium Publications; American Mathematical Society: Rhode Island, NY, USA, 1968; p. 39.
30. Zaikin, B.A.; Bogadarov, A.Y.; Kotov, A.F.; Poponov, P.V. Evaluation of coordinates of air target in a two-position range measurement radar. *Russ. Technol. J.* **2016**, *4*, 65–72.
31. Blahut, R.E. *Algebraic Codes for Data Transmission*; Cambridge University Press: Cambridge, UK, 2003.
32. Magomedov, S.G. Assessment of the impact of confounding factors in the performance information security. *Russ. Technol. J.* **2017**, *5*, 47–56.

33. Sigov, A.S.; Andrianova, E.G.; Zhukov, D.O.; Zykov, S.V.; Tarasov, I.E. Quantum informatics: overview of the main achievements. *Russ. Technol. J.* **2019**, *7*, 5–37. [CrossRef]
34. Shum, K.P.; Ren, X.; Wang, Y. Semigroups on semilattice and the constructions of generalized cryptogroups. *Southeast Asian Bull. Math.* **2014**, *38*, 719–730.

axioms

MDPI

Article

Smashed and Twisted Wreath Products of Metagroups

Sergey V. Ludkowski

Department of Applied Mathematics, MIREA—Russian Technological University, av. Vernadsky 78, 119454 Moscow, Russia; sludkowski@mail.ru

Received: 15 September 2019; Accepted: 30 October 2019; Published: 11 November 2019

Abstract: In this article, nonassociative metagroups are studied. Different types of smashed products and smashed twisted wreath products are scrutinized. Extensions of central metagroups are studied.

Keywords: metagroup; nonassociative; product; smashed; twisted wreath

MSC: 20N02; 20N05; 17A30; 17A60

1. Introduction

Nonassociative algebras compose a great area of algebra. In nonassociative algebra, noncommutative geometry, and quantum field theory, there frequently appear binary systems which are nonassociative generalizations of groups and related with loops, quasi-groups, Moufang loops, etc., (see References [1–4] and references therein). It was investigated and proved in the 20th century that a nontrivial geometry exists if and only if there exists a corresponding loop [1,5,6].

Octonions and generalized Cayley–Dickson algebras play very important roles in mathematics and quantum field theory [7–13]. Their structure and identities attract great attention. They are used not only in algebra and noncommutative geometry but also in noncommutative analysis, PDEs, particle physics, mathematical physics, the theory of Lie groups, algebras and their generalization, mathematical analysis, and operator theory and their applications in natural sciences including physics and quantum field theory (see References [7,11,12,14–19] and references therein).

A multiplicative law of their canonical bases is nonassociative and leads to a more general notion of a metagroup instead of a group [11,20,21]. The preposition "meta" is used to emphasize that such an algebraic object has properties milder than a group. By their axiomatics, metagroups satisfy the conditions of Equations (1)–(3) and rather mild relations (Equation (9)). They were used in References [20,21] for investigations of automorphisms, derivations, and cohomologies of nonassociative algebras. In the associative case, twisted and wreath products of groups are used for investigations not only in algebra but also in algebraic geometry, geometry, coding theory, and PDEs and their applications [22–25]. Twisted structures also naturally appear in investigations in the G-N theory of wave propagation of the components of displacement, stress, temperature distribution, and change in the volume fraction field in an isotropic homogeneous thermoelastic solid with voids subjected to thermal loading due to laser pulse [26].

In this article, nonassociative metagroups are studied. Necessary preliminary results on metagroups are described in Section 2. Quotient groups of metagroups are investigated in Theorem 1. Identities in metagroups established in Lemmas 1, 2, and 4 are applied in Sections 3 and 4.

Different types of smashed products of metagroups are investigated in Theorems 3 and 4. Besides them, direct products are also considered in Theorem 2. They provide large families of metagroups (see Remark 2).

In Section 4, smashed twisted wreath products of metagroups and particularly also of groups are scrutinized. It appears that, generally, they provide loops (see Theorem 5). If additional conditions are imposed, they give metagroups (see Theorem 6). Their metaisomorphisms are investigated in Theorem 7. In Theorem 8 and Corollary 2, smashed splitting extensions of nontrivial central metagroups are studied.

All main results of this paper are obtained for the first time. They can be used for further studies of binary systems, nonassociative algebra cohomologies, structure of nonassociative algebras, operator theory and spectral theory over Cayley–Dickson algebras, PDEs, noncommutative analysis, noncommutative geometry, mathematical physics, and their applications in the sciences (see also the conclusions).

2. Nonassociative Metagroups

To avoid misunderstandings, we give necessary definitions. A reader familiar with References [1,20,21] may skip Definition 1. For short, it will be written as a metagroup instead of a nonassociative metagroup.

Definition 1. *Let G be a set with a single-valued binary operation (multiplication) $G^2 \ni (a,b) \mapsto ab \in G$ defined on G satisfying the following conditions:*

$$\text{For each } a \text{ and } b \text{ in } G, \text{ there is a unique } x \in G \text{ with } ax = b \tag{1}$$

and a unique $y \in G$ exists satisfying $ya = b$, which are denoted by

$$x = a\backslash b = Div_l(a,b) \text{ and } y = b/a = Div_r(a,b) \text{ correspondingly,} \tag{2}$$

There exists a neutral (i.e., unit) element $e_G = e \in G$:

$$eg = ge = g \text{ for each } g \in G. \tag{3}$$

If the set G with the single-valued multiplication satisfies the conditions of Equations (1) and (2), then it is called a quasi-group. If the quasi-group G satisfies also the condition of Equation (3), then it is called an algebraic loop (or in short, a loop).

The set of all elements $h \in G$ commuting and associating with G is as follows:

$$Com(G) := \{a \in G : \forall b \in G,\ ab = ba\}, \tag{4}$$

$$N_l(G) := \{a \in G : \forall b \in G, \forall c \in G,\ (ab)c = a(bc)\}, \tag{5}$$

$$N_m(G) := \{a \in G : \forall b \in G, \forall c \in G,\ (ba)c = b(ac)\}, \tag{6}$$

$$N_r(G) := \{a \in G : \forall b \in G, \forall c \in G,\ (bc)a = b(ca)\}, \tag{7}$$

$$N(G) := N_l(G) \cap N_m(G) \cap N_r(G); \tag{8}$$

$$\mathcal{C}(G) := Com(G) \cap N(G) \text{ is called the center } \mathcal{C}(G) \text{ of } G.$$

We call G a metagroup if a set G possesses a single-valued binary operation and satisfies the conditions of Equations (1)–(3) and

$$(ab)c = t(a,b,c)a(bc) \tag{9}$$

for each a, b, and c in G, where $t(a,b,c) = t_G(a,b,c) \in \mathcal{C}(G)$. If G is a quasi-group satisfying the condition of Equation (9), then it will be called a strict quasi-group.

Then, the metagroup G will be called a central metagroup, if it satisfies also the following condition:

$$ab = t_2(a,b)ba \tag{10}$$

for each a and b in G, where $t_2(a,b) \in \mathcal{C}(G)$.

If H is a submetagroup (or a subloop) of the metagroup G (or the loop G) and $gH = Hg$ for each $g \in G$, then H will be called almost normal. If, in addition, $(gH)k = g(Hk)$ and $k(gH) = (kg)H$ for each g and k in G, then H will be called a normal submetagroup (or a normal subloop respectively).

Henceforward, notations $Inv_l(a) = Div_l(a,e)$ and $Inv_r(a) = Div_r(a,e)$ will be used.

Elements of a metagroup G will be denoted by small letters; subsets of G will be denoted by capital letters. If A and B are subsets in G, then $A - B$ means the difference of them: $A - B = \{a \in A : a \notin B\}$. Henceforward, maps and functions on metagroups are supposed to be single-valued unless otherwise specified.

Lemma 1. *If G is a metagroup, then for each a and b ∈ G, the following identities are fulfilled:*

$$b\backslash e = (e/b)t(e/b,b,b\backslash e) \tag{11}$$

$$(a\backslash e)b = (a\backslash b)t(e/a,a,a\backslash e)/t(e/a,a,a\backslash b); \tag{12}$$

$$b(e/a) = (b/a)t(b/a,a,a\backslash e)/t(e/a,a,a\backslash e). \tag{13}$$

Proof. The conditions of Equations (1)–(3) imply that

$$b(b\backslash a) = a,\ b\backslash(ba) = a; \tag{14}$$

$$(a/b)b = a,\ (ab)/b = a \tag{15}$$

for each a and b in G. Using the condition of Equation (9) and the identities of Equations (14) and (15), we deduce the following:

$$e/b = (e/b)(b(b\backslash e)) = (b\backslash e)/t(e/b,b,b\backslash e)$$

which leads to Equation (11).

Let $c = a\backslash b$; then, from the identities of Equations (11) and (14), it follows that

$$(a\backslash e)b = (e/a)t(e/a,a,a\backslash e)(ac) = ((e/a)a)(a\backslash b)t(e/a,a,a\backslash e)/t(e/a,a,a\backslash b)$$

which provides Equations (12).

Now, let $d = b/a$; then, the identities of Equations (11) and (15) imply that

$$b(e/a) = (da)(a\backslash e)/t(e/a,a,a\backslash e) = (b/a)t(b/a,a,a\backslash e)/t(e/a,a,a\backslash e)$$

which demonstrates Equation (13). □

Lemma 2. *Assume that G is a metagroup. Thenm for every a, a_1, a_2, and a_3 in G and p_1, p_2, and p_3 in $C(G)$, we have the following:*

$$t(p_1a_1,p_2a_2,p_3a_3) = t(a_1,a_2,a_3); \tag{16}$$

$$t(a,a\backslash e,a)t(a\backslash e,a,e/a) = e. \tag{17}$$

Proof. Since $(a_1a_2)a_3 = t(a_1,a_2,a_3)a_1(a_2a_3)$ and $t(a_1,a_2,a_3) \in C(G)$ for every a_1, a_2, a_3 in G, then

$$t(a_1,a_2,a_3) = ((a_1a_2)a_3)/(a_1(a_2a_3)). \tag{18}$$

Therefore, for every a_1, a_2, a_3 in G and p_1, p_2, and p_3 in $C(G)$, we infer the following:

$$t(p_1a_1,p_2a_2,p_3a_3) = (((p_1a_1)(p_2a_2))(p_3a_3))/((p_1a_1)((p_2a_2)(p_3a_3)))$$
$$= ((p_1p_2p_3)((a_1a_2)a_3))/((p_1p_2p_3)(a_1(a_2a_3))) = ((a_1a_2)a_3)/(a_1(a_2a_3)),$$

since

$$b/(pa) = p^{-1}b/a \text{ and } b/p = p\backslash b = bp^{-1} \tag{19}$$

For each $p \in \mathcal{C}(G)$, a and b in G, because $\mathcal{C}(G)$ is the commutative group. Thus, $t(p_1a_1, p_2a_2, p_3a_3) = t(a_1, a_2, a_3)$.

From the condition in Equation (9), Lemma 1, and the identity of Equation (16), it follows that

$$t(a, a\backslash e, a) = ((a(a\backslash e))a)/(a((a\backslash e)a)) = a/[at(e/a, a, a\backslash e)] = e/t(a\backslash e, a, e/a)$$

for each $a \in G$, implying Equation (17). □

Theorem 1. *If G is a metagroup and $\mathcal{C}_0$ is a subgroup in a center $\mathcal{C}(G)$ such that $t(a,b,c) \in \mathcal{C}_0$ for each a, b, and c in G, then its quotient $G/\mathcal{C}_0$ is a group.*

Proof. As traditionally, the following notation is used:

$$AB = \{x = ab : a \in A, b \in B\}, \tag{20}$$

$$Inv_l(A) = \{x = a\backslash e : a \in A\}, \tag{21}$$

$$Inv_r(A) = \{x = e/a : a \in A\} \tag{22}$$

for subsets A and B in G. Then from the conditions of Equations (4)–(8), it follows that, for each a, b, and c in G, the following identities take place:

$((a\mathcal{C}_0)(b\mathcal{C}_0))(c\mathcal{C}_0) = (a\mathcal{C}_0)((b\mathcal{C}_0)(c\mathcal{C}_0))$ and $a\mathcal{C}_0 = \mathcal{C}_0a$. Evidently $e\mathcal{C}_0 = \mathcal{C}_0$. In view of Lemmas 1 and 2 $(a\mathcal{C}_0)\backslash e = e/(a\mathcal{C}_0)$, consequently, for each $a\mathcal{C}_0 \in G/\mathcal{C}_0$ a unique inverse $(a\mathcal{C}_0)^{-1}$ exists. Thus the quotient $G/\mathcal{C}_0$ of G by $\mathcal{C}_0$ is a group. □

Lemma 3. *Let G be a metagroup, then $Inv_r(G)$ and $Inv_l(G)$ are metagroups.*

Proof. At first, we consider $Inv_r(G)$. Let a_1 and a_2 belong to G. Then, there are unique e/a_1 and e/a_2, since the map Inv_r is single-valued (see Definition 1). Since $Inv_r \circ Inv_l(a) = a$ and $Inv_l \circ Inv_r(a) = a$ for each $a \in G$, then $Inv_r : G \to G$ and $Inv_l : G \to G$ are bijective and surjective maps.

We put $\hat{a}_1 \circ \hat{a}_2 = (e/a_2)(e/a_1)$ for each a_1 and a_2 in G, where $\hat{a}_j = Inv_r(a_j)$ for each $j \in \{1, 2\}$. This provides a single-valued map from $Inv_r(G) \times Inv_r(G)$ into $Inv_r(G)$. Then, for each a, b, x, and y in G, the equations $\hat{a} \circ \hat{x} = \hat{b}$ and $\hat{y} \circ \hat{a} = \hat{b}$ are equivalent to $(e/x)(e/a) = e/b$ and $(e/a)(e/y) = e/b$, respectively. That is, $\hat{x} = (e/b)/(e/a)$ and $\hat{y} = (e/a)\backslash(e/b)$ are unique. On the other hand, $e/e = e$ and $\hat{e} \circ \hat{b} = e/b = \hat{b} \circ \hat{e} = \hat{b}$ for each $b \in G$.

Then, we infer the following:

$$\hat{a}_1 \circ (\hat{a}_2 \circ \hat{a}_3) = ((e/a_3)(e/a_2))(e/a_1) =$$
$$t_G(e/a_3, e/a_2, e/a_1)(e/a_3)((e/a_2)(e/a_1)) = t_G(\hat{a}_3, \hat{a}_2, \hat{a}_1)(\hat{a}_1 \circ \hat{a}_2) \circ \hat{a}_3,$$

Consequently, $t_{Inv_r(G)}(\hat{a}_1, \hat{a}_2, \hat{a}_3) = e/t_G(\hat{a}_3, \hat{a}_2, \hat{a}_1)$. Evidently, $Inv_r(\mathcal{C}(G)) = \mathcal{C}(G)$ and $\mathcal{C}(Inv_r(G)) = \mathcal{C}(G)$. Thus, the conditions of Equations (1)–(3) and (9) are satisfied for $Inv_r(G)$.

Similarly, putting $Inv_l(a_j) = \check{a}_j$ and $\check{a}_1 \circ \check{a}_2 = (a_2\backslash e)(a_1\backslash e)$ for each $a_j \in G$ and $j \in \{1, 2, 3\}$, the conditions of Equations (1)–(3) and (9) are verified for $Inv_l(G)$. □

Lemma 4. *Assume that G is a metagroup and that $a \in G$, $b \in G$, and $c \in G$. Then*

$$e/(ab) = (e/b)(e/a)t(e/a, a, b)/t(e/b, e/a, ab) \tag{23}$$

and

$$(ab)\backslash e = (b\backslash e)(a\backslash e)t(ab, b \setminus e, a\backslash e)/t(a, b, b\backslash e). \tag{24}$$

$$(a/(bc) = ((a/c)/b)t(a/(bc), b, c), \tag{25}$$

$$(bc)\backslash a = (c\backslash(b\backslash a))/t(b, c, (bc) \setminus a). \tag{26}$$

Proof. From Equations (9) and (15), we deduce that

$((e/b)(e/a))(ab) = t(e/b, e/a, ab)(e/b)((e/a)(ab)) = t(e/b, e/a, ab)/t(e/a, a, b)$, which implies Equation (23). Then, from Equations (9) and (14), we infer the following:

$$(ab)((b\backslash e)(a\backslash e)) = ((ab)(b\backslash e))(a\backslash e)/t(ab, b\backslash e, a\backslash e) = t(a, b, b\backslash e)/t(ab, b\backslash e, a\backslash e)$$

which implies Equation (24).

Utilizing Equations (14) and (9), we get $b(c((bc) \setminus a)) = a/t(b, c, (bc)\backslash a)$; hence, $c((bc)\backslash a) = (b \setminus a)/t(b, c, (bc)\backslash a)$, implying Equation (26).

Equations (15) and (9) imply that $((a/(bc))b)c = t(a/(bc), b, c)a$; consequently, $(a/c)t(a/(bc), b, c) = (a/(bc))b$, and hence,

$$((a/c)/b)t(a/(bc), b, c) = a/(bc).$$

□

3. Smashed Products and Smashed Twisted Products of Metagroups

Theorem 2. *Let G_j be a family of metagroups (see Definition 1), where $j \in J$, J is a set. Then, their direct product $G = \prod_{j \in J} G_j$ is a metagroup and*

$$\mathcal{C}(G) = \prod_{j \in J} \mathcal{C}(G_j). \tag{27}$$

Proof. It is given in Theorem 8 in Reference [21]. □

Remark 1. *Let A and B be two metagroups, and let $\mathcal{C}$ be a commutative group such that*

$$\mathcal{C}_m(A) \hookrightarrow \mathcal{C}, \mathcal{C}_m(B) \hookrightarrow \mathcal{C}, \mathcal{C} \hookrightarrow \mathcal{C}(A) \text{ and } \mathcal{C} \hookrightarrow \mathcal{C}(B), \tag{28}$$

where $\mathcal{C}_m(A)$ denotes a minimal subgroup in $\mathcal{C}(A)$ containing $t_A(a, b, c)$ for every a, b, and c in A.

Using direct products, it is always possible to extend either A or B to get such a case. In particular, either A or B may be a group. On $A \times B$, an equivalence relation Ξ is considered such that

$$(\gamma v, b)\Xi(v, \gamma b) \text{ and } (\gamma v, b)\Xi\gamma(v, b) \text{ and } (\gamma v, b)\Xi(v, b)\gamma \tag{29}$$

for every v in A, b in B, and γ in $\mathcal{C}$.

$$\text{Let } \phi : A \to \mathcal{A}(B) \text{ be a single-valued mapping,} \tag{30}$$

where $\mathcal{A}(B)$ denotes a family of all bijective surjective single-valued mappings of B onto B subjected to the conditions of Equations (31)–(34) given below. If $a \in A$ and $b \in B$, then it will be written shortly b^a instead of $\phi(a)b$, where $\phi(a) : B \to B$. Let also

$$\eta_{A,B,\phi} : A \times A \times B \to \mathcal{C}, \kappa_{A,B,\phi} : A \times B \times B \to \mathcal{C}$$

and

$$\xi_{A,B,\phi} : ((A \times B)/\Xi) \times ((A \times B)/\Xi) \to \mathcal{C}$$

be single-valued mappings written shortly as η, κ, *and* ξ *correspondingly such that*

$$(b^u)^v = b^{vu}\eta(v,u,b),\ e^u = e,\ b^e = b; \tag{31}$$

$$\eta(v,u,\gamma b) = \eta(v,u,b); \tag{32}$$

$$(cb)^u = c^u b^u \kappa(u,c,b); \tag{33}$$

$$\kappa(u,\gamma c,b) = \kappa(u,c,\gamma b) = \kappa(u,c,b) \tag{34}$$

and $\kappa(u,\gamma,b) = \kappa(u,b,\gamma) = e$;

$$\xi((\gamma u,c),(v,b)) = \xi((u,c),(\gamma v,b)) = \xi((u,c),(v,b))$$

and

$$\xi((\gamma,e),(v,b)) = e \text{ and } \xi((u,c),(\gamma,e)) = e \tag{35}$$

for every u *and* v *in* A, b, *and* c *in* B, γ *in* $\mathcal{C}$, *where* e *denotes the neutral element in* $\mathcal{C}$ *and in* A *and* B.

We put

$$(a_1,b_1)(a_2,b_2) = (a_1a_2, \xi((a_1,b_1),(a_2,b_2))b_1b_2^{a_1}) \tag{36}$$

for each of a_1 *and* a_2 *in* A *and of* b_1 *and* b_2 *in* B.

The Cartesian product $A \times B$ *supplied with such a binary operation of Equation* (36) *will be denoted by* $A \bigotimes^{\phi,\eta,\kappa,\xi} B$.

Then, we put

$$(a_1,b_1) \star (a_2,b_2) = (a_1a_2, \xi((a_1,b_1),(a_2,b_2))b_2^{a_1}b_1) \tag{37}$$

for each of a_1 *and* a_2 *in* A *and of* b_1 *and* b_2 *in* B.

The Cartesian product $A \times B$ *supplied with a binary operation of Equation* (37) *will be denoted by* $A \star^{\phi,\eta,\kappa,\xi} B$.

Theorem 3. *Let the conditions of Remark 1 be fulfilled. Then, the Cartesian product* $A \times B$ *supplied with a binary operation of Equation* (36) *is a metagroup. Moreover, there are embeddings of* A *and* B *into* $A \bigotimes^{\phi,\eta,\kappa,\xi} B = C_1$ *such that* B *is an almost normal submetagroup in* C_1. *If in addition* $C_m(C_1) \subseteq C_m(B) \subseteq \mathcal{C}$, *then* B *is a normal submetagroup.*

Proof. The first part of this theorem was proven in Theorem 9 in Reference [21]. Naturally, A is embedded into C_1 as $\{(a,e) : a \in A\}$ and B is embedded into C_1 as $\{(e,b) : b \in B\}$. Let $a \in A$ and $b_0 \in B$; then, $(a,b_0)B = \{(a, \xi((a,b_0),(e,b))b_0b^a) : b \in B\}$ and $B(a,b_0) = \{(a, \xi((e,b),(a,b_0))bb_0) : b \in B\}$, since $b_0^e = b_0$ by (31). From $B^a = B$, $b_0B = B$, $Bb_0 = B$, $\mathcal{C} \subset C(B)$ and Equations (30) and (35), it follows that $(a,b_0)B = B(a,b_0)$, where $B^a = \{b^a : b \in B\}$. Thus, B is an almost normal submetagroup in C_1 (see Definition 1). If in addition $C_m(C_1) \subseteq C_m(B) \subseteq \mathcal{C}$, then evidently B is a normal submetagroup (see also the condition of Equation (29)), since $t_{C_1}(g,b,h) \in C_m(C_1)$ and $t_{C_1}(h,g,b) \in C_m(C_1)$ for each g and h in G, $b \in H$. □

Theorem 4. *Suppose that the conditions of Remark 1 are satisfied. Then, the Cartesian product* $A \times B$ *supplied with a binary operation of Equation* (37) *is a metagroup. Moreover, there exist embeddings of* A *and* B *into* $A \star^{\phi,\eta,\kappa,\xi} B = C_2$ *such that* B *is an almost normal submetagroup in* C_2. *If additionally* $C_m(C_2) \subseteq C_m(B) \subseteq \mathcal{C}$, *then* B *is a normal submetagroup.*

Proof. The conditions of Remark 1 imply that the binary operation of Equation (37) is single-valued.

We consider the following formulas:

$I_1 = ((a_1, b_1) \star (a_2, b_2)) \star (a_3, b_3)$ and $I_2 = (a_1, b_1) \star ((a_2, b_2) \star (a_3, b_3))$, where a_1, a_2, and a_3 are in A and where b_1, b_2, and b_3 are in B. Utilizing Equations (31)–(35) and (37), we get the following:

$$I_1 = ((a_1a_2)a_3, \xi((a_1a_2, b_2^{a_1}b_1), (a_3, b_3))\xi((a_1, b_1), (a_2, b_2))b_3^{a_1a_2}(b_2^{a_1}b_1))$$

and

$$I_2 = (a_1(a_2a_3), \xi((a_1, b_1), (a_2a_3, b_3^{a_2}b_2))[\xi((a_2, b_2), (a_3, b_3))]^{a_1}(b_3^{a_1a_2}b_2^{a_1})b_1\kappa(a_1, b_3^{a_2}, b_2)\eta(a_1, a_2, b_3)).$$

Therefore

$$I_1 = t((a_1, b_1), (a_2, b_2), (a_3, b_3))I_2 \tag{38}$$

with

$$\begin{aligned} t((a_1, b_1), (a_2, b_2), (a_3, b_3)) &= t_A(a_1, a_2, a_3)\xi((a_1, b_1), (a_2, b_2)) \\ \xi((a_1a_2, b_2^{a_1}b_1), (a_3, b_3))/\{t_B(b_3^{a_1a_2}, b_2^{a_1}, b_1) \\ \xi((a_1, b_1), (a_2a_3, b_3^{a_2}b_2))[\xi((a_2, b_2), (a_3, b_3))]^{a_1}\kappa(a_1, b_3^{a_2}, b_2)\eta(a_1, a_2, b_3))\}, \end{aligned} \tag{39}$$

Consequently, $t((a_1, b_1), (a_2, b_2), (a_3, b_3)) \in \mathcal{C}$ for each $a_j \in A$, $b_j \in B$, $j \in \{1, 2, 3\}$. We denote

$$t((a_1, b_1), (a_2, b_2), (a_3, b_3))$$

in more details by

$$t_{A\star^{\phi,\eta,\kappa,\xi}B}((a_1, b_1), (a_2, b_2), (a_3, b_3))$$

(see Equation (39)).

Evidently, Equation (3) is a consequence of Equations (35) and (37).

Note that, if $\gamma \in \mathcal{C}$, then

$$\begin{aligned} \gamma((a_1, b_1) \star (a_2, b_2)) = (\gamma a_1a_2, \xi((a_1, b_1), (a_2, b_2))b_2^{a_1}b_1) = \\ (a_1a_2, b_2^{a_1}b_1)\gamma\xi((a_1, b_1), (a_2, b_2)) = ((a_1, b_1) \star (a_2, b_2))\gamma. \end{aligned}$$

Therefore, $\gamma \in \mathcal{C}(A \star^{\phi,\eta,\kappa,\xi} B)$. Consequently, $\mathcal{C} \subseteq \mathcal{C}(A \star^{\psi,\eta,\kappa,\xi} B)$.

Then, we seek a solution of the following equation:

$$(a_1, b_1) \star (a, b) = (e, e), \tag{40}$$

where $a \in A, b \in B$.

From Equations (2) and (37), it follows that

$$a_1 = e/a \tag{41}$$

Consequently, $\xi((e/a, b_1), (a, b))b^{(e/a)}b_1 = e$. Therefore, Equations (1) and (35) imply that

$$b_1 = [\xi((e/a, b^{(e/a)}), (a, b))b^{(e/a)}]\backslash e. \tag{42}$$

Thus, $a_1 \in A$ and $b_1 \in B$ prescribed by Equations (41) and (42) provide a unique solution of Equation (40).

Analogously for the following equation

$$(a, b)(a_2, b_2) = (e, e), \tag{43}$$

where $a \in A$, $b \in B$, we deduce that

$$a_2 = a \backslash e. \tag{44}$$

Consequently, $\xi((a,b),(a\backslash e,b_2))b_2^a b = e$, and hence, $b_2^a = e/[\xi((a,b),(a\backslash e,b_2))b]$. From Equations (31) and (32), it follows that $(b_2^a)^{e/a} = \eta(e/a,a,b_2)b_2$; consequently,

$$b_2 = (e/b)^{e/a}/\{[(\xi((a,b),(a\backslash e,(e/b)^{e/a}))]^{e/a}\eta(e/a,a,(e/b)^{e/a})\}. \tag{45}$$

Thus, a unique solution of Equation (43) is given by Equations (44) and (45).

Then, we have $(a_1,b_1) = (e,e)/(a,b)$ and $(a_2,b_2) = (a,b)\backslash(e,e)$ and get the following:

$$(a,b) \setminus (c,d) = ((a,b)\backslash(e,e))(c,d)$$

$$t((e,e)/(a,b),(a,b),((a,b)\backslash(e,e))(c,d))/t((e,e)/(a,b),(a,b),(a,b)\backslash(e,e)); \tag{46}$$

$$(c,d)/(a,b) = (c,d)((e,e)/(a,b))$$
$$t((e,e)/(a,b),(a,b),(a,b)\backslash(e,e))/t((c,d)(e/(a,b)),(a,b),(a,b)\backslash(e,e)) \tag{47}$$

and $e_G = (e,e)$, where $G = A \star^{\phi,\eta,\kappa,\xi} B$. This means that the properties of Equations (1)–(3) and (9) are fulfilled for $A \star^{\phi,\eta,\kappa,\xi} B$.

Evidently, there are embeddings of A and B into C_2 as (A,e) and (e,B), respectively. Suppose that $a \in A$ and $b_0 \in B$, then

$(a,b_0) \star B = \{(a,\xi((a,b_0),(e,b))b^a b_0) : b \in B\}$ and

$B \star (a,b_0) = \{(a,\xi((e,b),(a,b_0))b_0 b) : b \in B\}$.

Therefore, $(a,b_0) \star B = B \star (a,b_0)$ by the conditions of Equations (30) and (35), since $B^a = B$ and $\mathcal{C} \subset \mathcal{C}(B)$. Thus, B is an almost normal submetagroup in C_2 (see Definition 1). If additionally $\mathcal{C}_m(C_2) \subseteq \mathcal{C}_m(B) \subseteq \mathcal{C}$, then apparently B is a normal submetagroup (see also the condition of Equation (29)), since $t_{C_2}(g,b,h) \in \mathcal{C}_m(C_2)$ and $t_{C_2}(h,g,b) \in \mathcal{C}_m(C_2)$ for each g and h in G, $b \in B$. □

Definition 2. *We call the metagroup $A \bigotimes^{\phi,\eta,\kappa,\xi} B$ provided by Theorem 3 (or $A \star^{\phi,\eta,\kappa,\xi} B$ by Theorem 4) a smashed product (or a smashed twisted product correspondingly) of metagroups A and B with smashing factors ϕ, η, κ, and ξ.*

Remark 2. *From Theorems 2–4, it follows that, taking nontrivial η, κ, and ξ and starting even from groups with nontrivial $\mathcal{C}(G_j)$ or $\mathcal{C}(A)$, it is possible to construct new metagroups with nontrivial $\mathcal{C}(G)$ and ranges $t_G(G,G,G)$ of t_G that may be infinite.*

With suitable smashing factors ϕ, η, κ, and ξ and with nontrivial metagroups or groups A and B, it is easy to get examples of metagroups in which $e/a \neq a\backslash e$ for an infinite family of elements a in $A \bigotimes^{\phi,\eta,\kappa,\xi} B$ or in $A \star^{\phi,\eta,\kappa,\xi} B$. Evidently, smashed products and smashed twisted products (see Definition 2) are nonassociative generalizations of semidirect products. Combining Theorems 3 and 4 with Lemmas 3 and 4 provides other types of smashed products by taking $\hat{b}_1 \circ \hat{b}_2^{a_1}$ instead of $b_1 b_2^{a_1}$ or $\check{b}_2^{a_1} \circ \check{b}_1$ instead of $b_2^{a_1} b_1$ on the right sides of Equations (36) and (37), correspondingly, etc.

4. Smashed Twisted Wreath Products of Metagroups

Lemma 5. *Let D be a metagroup and A be a submetagroup in D. Then, there exists a subset V in D such that D is a disjoint union of vA, where $v \in V$, that is,*

$$D = \bigcup_{v \in V} vA \tag{48}$$

and

$$(\forall v_1 \in V,\ \forall v_2 \in V,\ v_1 \neq v_2) \Rightarrow (v_1 A \cap v_2 A) = \emptyset). \quad (49)$$

Proof. The cases $A = \{e\}$ and $A = D$ are trivial. Let $A \neq \{e\}$ and $A \neq D$, and let $\mathcal{C}(D)$ be a center of D. From the conditions of Equations (4)–(8), it follows that $z \in \mathcal{C}(D) \cap A$ implies $z \in \mathcal{C}(A)$.

Assume that $b \in D$ and $z \in \mathcal{C}(D)$ are such that $zbA \cap bA \neq \emptyset$. It is equivalent to $(\exists s_1 \in A, \exists s_2 \in A, zbs_1 = bs_2)$. From Equation (14), it follows that $(zbs_1 = bs_2) \Leftrightarrow (zs_1 = s_2) \Leftrightarrow (z = s_2/s_1 \in A)$ because $z \in \mathcal{C}(D)$. Thus,

$$(\exists b \in D,\ \exists z \in \mathcal{C}(D),\ zbA \cap bA = \emptyset) \Leftrightarrow (\exists b \in D,\ \exists z \in \mathcal{C}(D) - A). \quad (50)$$

Suppose now that $b_1 \in D$, $b_2 \in D$ and $b_1 A \cap b_2 A \neq \emptyset$. This is equivalent to $(\exists s_1 \in A, \exists s_2 \in A, b_1 s_1 = b_2 s_2)$. By the identity of Equation (15), the latter is equivalent to $b_1 = (b_2 s_2)/s_1$. On the other hand,

$$(b_2s_2)/s_1 = (b_2s_2)(e/s_1)t(e/s_1, s_1, s_1 \setminus e)/t((b_2s_2)/s_1, s_1, s_1 \setminus e)$$
$$= b_2(s_2(e/s_1))t(b_2, s_2, e/s_1)t(e/s_1, s_1, s_1 \setminus e)/t((b_2s_2)/s_1, s_1, s_1 \setminus e)$$

by Equations (9), (13), and (15). Together with (50) this gives the equivalence:

$$(\exists b_1 \in D,\ \exists b_2 \in D,\ b_1 A \cap b_2 A \neq \emptyset) \Leftrightarrow (\exists b_1 \in D,\ \exists b_2 \in D,\ \exists s \in A, \exists z \in \mathcal{C}(D) - A,\ b_1 = zb_2 s). \quad (51)$$

Let Y be a family of subsets K in D such that $k_1 A \cap k_2 A = \emptyset$ for each $k_1 \neq k_2$ in K. Let Y be directed by inclusion. Then, $Y \neq \emptyset$, since $A \subset D$ and $A \neq D$. Therefore, from Equations (50) and (51) and the Kuratowski-Zorn lemma (see Reference [27]), the assertion of this lemma follows, since a maximal element V in Y gives Equations (48) and (49). □

Definition 3. *A set V from Lemma 5 is called a right transversal (or complete set of right coset representatives) of A in D.*

The following corollary is an immediate consequence of Lemma 5.

Corollary 1. *Let D be a metagroup, A be a submetagroup in D, and V be a right transversal of A in D. Then,*

$$\forall a \in D,\ \exists_1 s \in A,\ \exists_1 b \in V,\ a = sb \text{ for a given triple } (A, D, V). \quad (52)$$

Remark 3. *We denote b in the decomposition of Equation (52) by $b = \tau(a) = a^\tau$ and $s = \psi(a) = a^\psi$, where τ and ψ are the shortened notations of $\tau_{A,D,V}$ and $\psi_{A,D,V}$, respectively. That is, there are single-valued maps*

$$\tau : D \to V \text{ and } \psi : D \to A. \quad (53)$$

These maps are idempotent $\tau(\tau(a)) = \tau(a)$ and $\psi(\psi(a)) = \psi(a)$ for each $a \in D$.

$$\text{If } b = a^\tau, \text{ then we denote } e/b \text{ by } a^{e/\tau} \text{ and } b\setminus e \text{ by } a^{\tau\setminus e}. \quad (54)$$

According to Equation (2), $s = a/b$; hence, $a^\psi = a/a^\tau$. From Equation (13), it follows that $a/b = a(e/b)t(e/b, b, b \setminus e)/t(a/b, b, b\setminus e)$; consequently, by Lemma 2,

$$s = aa^{e/\tau}t(a^{e/\tau}, a^\tau, a^{\tau\setminus e})/t(aa^{e/\tau}, a^\tau, a^{\tau\setminus e}). \quad (55)$$

Notice that the metagroup need not be power-associative. Then, e/s and $s\setminus e$ can be calculated with the help of the identity of Equation (11). Suppose that a and y belong to D, $s = a^\psi$, $b = a^\tau$, $s_2 = y^\psi$, and $b_2 = y^\tau$. Then, $(a^\tau y) = b(s_2 b_2)$. According to Equation (52) there exists a unique decomposition $b(s_2b_2) = s_3b_3$, where $s_3 \in A$, $b_3 \in V$; hence, $(a^\tau y)^\tau = b_3$. On the other hand, by Equation (9) $ay = s(b(s_2b_2))t(s, b, y) =$

$(ss_3)b_3t(s,b,y)/t(s,s_3,b_3)$. *We denote a subgroup* $\mathcal{C}(D) \cap A$ *in* $\mathcal{C}(D)$ *by* $\mathcal{C}_A(D)$ *or shortly* $\mathcal{C}_A$, *when D is specified. From Lemma 2 and Equation* (51), *it follows that*

$$\mathcal{C}(D)^\tau \text{ is isomorphic with } \mathcal{C}(D)/\mathcal{C}_A, \tag{56}$$

where $\mathcal{C}(D)^\tau = \{a^\tau : a \in \mathcal{C}(D)\}$.

Let $\mathcal{C}_m(A)$ *be a minimal subgroup in* $\mathcal{C}(A)$ *generated by a set* $\{t_A(a,b,c) : a \in A, b \in A, c \in A\}$. *From Equation* (9), *it follows that* $\mathcal{C}_m(A) \subset \mathcal{C}_A(D)$ *and* $A\mathcal{C}(D)$ *is a submetagroup in D. By virtue of Theorem 1,* $(A\mathcal{C}(D))/\mathcal{C}_A(D)$ *and* $A/\mathcal{C}_A(D)$ *are groups such that* $A/\mathcal{C}_A(D) \hookrightarrow (A\mathcal{C}(D))/\mathcal{C}_A(D)$. *For each* $d \in D$, *there exists a unique decomposition*

$$d = d^\psi d^\tau \tag{57}$$

by Equation (53). *Take in particular* $\gamma \in \mathcal{C}(D)$; *then,* $\gamma = \gamma^\psi \gamma^\tau$, *where* $\gamma^\psi \in \mathcal{C}_A(D)$, $\gamma^\tau \in V$. *Therefore,* $\mathcal{C}(D)/\mathcal{C}_A(D) \subset V$ *and there exists a subset* V_0 *in V such that* $(\mathcal{C}(D)/\mathcal{C}_A(D))V_0 = V$, *since* $\mathcal{C}(D)/\mathcal{C}_A(D)$ *is a subgroup in* $(A\mathcal{C}(D))/\mathcal{C}_A(D)$ *(see Equation* (56)*). Equation* (57) *implies that* $(d^\tau)^\psi = e$ *and* $(d^\psi)^\tau = e$ *for each* $d \in D$. *Using this, we subsequently deduce that*

$$(d^\psi \gamma)^\psi = d^\psi \gamma^\psi, \tag{58}$$

$$(d^\psi \gamma)^\tau = \gamma^\tau, \tag{59}$$

$$(d^\tau \gamma)^\psi = \gamma^\psi, \tag{60}$$

$$(d^\tau \gamma)^\tau = d^\tau \gamma^\tau \tag{61}$$

for each $d \in D$ *and* $\gamma \in \mathcal{C}(D)$. *Hence,*

$(d\gamma) = (d\gamma)^\psi (d\gamma)^\tau = (d^\psi d^\tau)(\gamma^\psi \gamma^\tau) = (d^\psi \gamma^\psi)(d^\tau \gamma^\tau) = (d^\psi \gamma)^\psi (d^\tau \gamma)^\tau$,

where $d^\psi \gamma^\psi \in A$ *and* $d^\tau \gamma^\tau \in V$. *From a uniqueness of this representation, it follows that*

$$(d\gamma)^\psi = d^\psi \gamma^\psi \tag{62}$$

and

$$(d\gamma)^\tau = d^\tau \gamma^\tau \text{ for each } d \in D \text{ and } \gamma \in \mathcal{C}(D). \tag{63}$$

Using Equation (63) *we infer that*

$$(a^\tau y)^\tau = (ay)^\tau [t_D(a^\psi, (a^\tau y)^\psi, (ay)^\tau)/t_D(a^\psi, a^\tau, y)]^\tau. \tag{64}$$

On the other hand, if $\gamma \in \mathcal{C}(D)$, *then* $\gamma^\psi = \gamma/\gamma^\tau$ *and Equations* (64) *and* (60) *imply particularly that*

$$(a^\tau \gamma)^\tau = (a\gamma)^\tau \text{ for each } a \in D \text{ and } \gamma \in \mathcal{C}(D), \tag{65}$$

since $t_D(a,d,\gamma) = e$ *for each a and d in D and* $\gamma \in \mathcal{C}(D)$. *Then, from* $s = a^\psi$, $a^\tau y = s_3 b_3$, *it follows that* $a^\psi (a^\tau y)^\psi = ss_3$ *and* $(ay)^\psi = [(ss_3)b_3 t_D(s,b,y)/t_D(s,s_3,b_3)]^\psi$; *consequently, by Lemma 2 and Equation* (62),

$$a^\psi (a^\tau y)^\psi = (ay)^\psi [t_D(a^\psi, (a^\tau y)^\psi, (ay)^\tau)/t_D(a^\psi, a^\tau, y)]^\psi \tag{66}$$

for each a and y in D. Particularly,

$$a^\psi (a^\tau \gamma)^\psi = (a\gamma)^\psi \text{ for each } a \in D \text{ and } \gamma \in \mathcal{C}(D). \tag{67}$$

From Equations (64) *and* (65), *it follows that the metagroup D acts on V transitively by right shift operators* R_y, *where* $R_y a = ay$ *for each a and y in D. Therefore, we put*

$$(a^\tau)^{[c]} := (a^\tau c)^\tau \text{ for each } a \text{ and } c \text{ in } D. \tag{68}$$

Then from Equations (64), (65), (68), and (9) and Lemma 2, we deduce that, for each a, c, and d in D

$$(a^\tau)^{[cd]} = ((a^\tau)^{[c]})^{[d]}[t_D((a^\tau c)^\psi, (a^\tau c)^\tau, d)/(t_D((a^\tau c)^\psi, ((a^\tau c)^\tau d)^\psi, ((a^\tau c)d)^\tau)t_D(a^\tau, c, d))]^\tau. \quad (69)$$

In particular, $(a^\tau)^{[e]} = a^\tau$ *for each* $a \in D$. *Next, we put* $e^\tau = b_*$. *It is convenient to choose* $b_* = e$. *Hence,* $b_*^{[s]} = (e^\tau)^{[s]} = (e^\tau s)^\tau = s^\tau = e = e^\tau$ *for each* $s \in A$. *Thus, the submetagroup A is the stabilizer of e and Equation (68) implies that*

$$e^{[s]} = e \text{ and } e^{[q]} = q \text{ for each } s \in A \text{ and } q \in V. \quad (70)$$

Remark 4. *Let B and D be metagroups, A be a submetagroup in D, and V be a right transversal of A in D. Let also the conditions of Equations (28)–(35) be satisfied for A and B. By Theorem 2, there exists a metagroup*

$$F = B^V, \text{ where } B^V = \prod_{v \in V} B_v,\ B_v = B \text{ for each } v \in V. \quad (71)$$

It contains a submetagroup
$F^* = \{f \in F : card(\sigma(f)) < \aleph_0\}$,
where $\sigma(f) = \{v \in V : f(v) \neq e\}$ *is a support of* $f \in F$ *and* $card(\Omega)$ *denotes the cardinality of a set* Ω.
Let $T_h f = f^h$ *for each* $f \in F$ *and* $h : V \to A$. *We put*
$\hat{S}_d(T_h f J) = T_{hS_d^{-1}} f S_d J$,
where $J : V \times F \to B$, $J(f, v) = fJv$, $S_d J v = J v^{[d \backslash e]}$ *for each* $d \in D$, $f \in F$ *and* $v \in V$. *Then, for each* $f \in F$, $d \in D$ *we put*

$$f^{\{d\}} = \hat{S}_d(T_{g_d} f E), \quad (72)$$

where

$$s(d, v) = e/(v/d)^\psi,\ g_d(v) = s(d, v),$$

$$fEv = f(v) \text{ for each } v \in V \quad (73)$$

(see also Equations (52) and (68)). Hence,

$$f^{\{e\}} = f, \quad (74)$$

since $v^{e \backslash e} = v$ *and* $s(e, v) = e$.

Lemma 6. *Let the conditions of Remark 4 be satisfied. Then, for each of f and* f_1 *in F and of d and* d_1 *in D,* $v \in V$,

$$(ff_1)^{\{d\}}(v) = \kappa(s(d, v), f(v^{[d \backslash e]}), f_1(v^{[d \backslash e]}))f^{\{d\}}(v)f_1^{\{d\}}(v) \quad (75)$$

and

$$f^{\{dd_1\}}(v) = \{[(f^{\{d_1\}})^{\{d\}}]^{w_2(d, d_1, v)}(vw_1(d, d_1, v))\}w_3(d, d_1, v), \quad (76)$$

where $w_j = w_j(d, d_1, v) \in \mathcal{C}(D)$, $j \in \{1, 2, 3\}$, $w_1^\tau = w_1$.

Proof. Equations (72) and (33) imply the identity of Equation (75).
Let $v \in V$, d and d_1 belong to D, and $f \in F$; then, from Equations (72) and (73), it follows that

$$f^{\{dd_1\}}(v) = f^{s(dd_1, v)}(v^{[(dd_1) \backslash e]}) \quad (77)$$

and

$$(f^{\{d_1\}})^{\{d\}}(v) = (f^{s(d_1, v)})^{s(d, v^{[d_1 \backslash e]})}((v^{[d_1 \backslash e]})^{[d \backslash e]}). \quad (78)$$

From Equations (24), (69), (58), (61), (11), and (13) and Lemma 2, we deduce that

$$(dd_1) \backslash e = (d_1 \backslash e)(d \backslash e)t_D(dd_1, d_1 \backslash e, d \backslash e)/t_D(d, d_1, d_1 \backslash e)$$

and

$$v^{[(dd_1)\backslash e]} = (v^{[d_1\backslash e]})^{[d\backslash e]} w_1(d, d_1, v), \tag{79}$$

where $w_1(d, d_1, v) = \gamma^\tau$, where

$$\gamma = t_D(dd_1, d_1\backslash e, d \setminus e) t_D((v/d_1)^\psi, (v/d_1)^\tau, d\backslash e)$$
$$/[t_D(d, d_1, d_1\backslash e) t_D(v, e/d_1, e/d) t_D((v/d_1)^\psi, ((v/d_1)^\tau/d)^\psi, (v/(dd_1))^\tau)].$$

Then Equations (73), (66), (25), (13), and (62) and Lemma 2 imply that

$$s(dd_1, v) = e/([(v/d_1)(e/d)]^\psi \gamma_1^\psi, \tag{80}$$

where $\gamma_1 = t_D(v/(dd_1), d, d_1) t_D(e/d, d, d \setminus e)/t_D(v/(dd_1), d, d\backslash e)$ by (13), (25) and Lemma 2;

$$[(v/d_1)(e/d)]^\psi = (v/d_1)^\psi [(v/d_1)^\tau (e/d)]^\psi \gamma_2^\psi, \tag{81}$$

where

$$\gamma_2 = t_D((v/d_1)^\psi, (v/d_1)^\tau, e/d)/t_D((v/d_1)^\psi, ((v/d_1)^\tau (e/d))^\psi, ((v/d_1)(e/d))^\tau).$$

Note that

$$s(dd_1, v) = (e/[(v/d_1)^\tau (e/d)]^\psi)(e/(v/d_1)^\psi)\gamma_3/\{\gamma_1^\psi \gamma_2^\psi\} \tag{82}$$

by Equations (81) and (23), where

$$\gamma_3 = t_D(e/(v/d_1)^\psi, (v/d_1)^\psi, [(v/d_1)^\tau (e/d)]^\psi)$$
$$/t_D(e/[(v/d_1)^\tau (e/d)]^\psi, e/(v/d_1)^\psi, (v/d_1)^\psi [(v/d_1)^\tau (e/d)]^\psi).$$

Then,

$$(v/d_1)^\tau = [v(d_1\backslash e)]^\tau \gamma_4^\tau = v^{[d_1\backslash e]} \gamma_4^\tau$$

by Equations (11), (13), (63), and (68), where $\gamma_4 \in \mathcal{C}_m(D)$. Hence,

$$[(v/d_1)^\tau (e/d)]^\psi = [v^{[d_1\backslash e]}(e/d)]^\psi \gamma_5^\psi,$$

since

$$(\gamma_4^\tau)^\psi = e, \tag{83}$$

where $\gamma_5 = t_D(v^{[d_1\backslash e]}/d, d, d\backslash e)/t_D(e/d, d, d \setminus e)$.

Thus, the identities of Equations (80)–(83) imply that

$$s(dd_1, v) = s(d, v^{[d_1\backslash e]}) s(d_1, v) w_2(d, d_1, v), \tag{84}$$

where

$$w_2(d, d_1, v) = \gamma_3/(\gamma_1^\psi \gamma_2^\psi \gamma_5^\psi),\ w_2(d, d_1, v) \in \mathcal{C}(D).$$

By Lemmas 1 and 2 and Equation (73), representations of γ_j simplify:

$$\gamma_2 = t_D(e/s(d_1, v), v^{[d_1\backslash e]}, e/d)/t_D(e/s(d_1, v), e/s(d, v^{[d_1\backslash e]}), v^{[(dd_1)\backslash e]}),$$
$$\gamma_3 = t_D(s(d_1, v), e/s(d_1, v), e/s(d, v^{[d_1\backslash e]}))$$
$$/t_D(s(d, v^{[d_1\backslash e]}), s(d_1, v), e/(s(d, v^{[d_1\backslash e]}) s(d_1, v))).$$

Therefore, $w_2(d, d_1, v) \in \mathcal{C}(D) \cap A$ for each d and d_1 in D and $v \in V$, since $s(d, d_1, v)$, $s(d, v^{[d_1\backslash e]})$, and $s(d_1, v)$ belong to A. Then, from Equations (77), (84), (31), (32), we infer that

$$f^{\{dd_1\}}(v) = \{[(f^{s(d_1,v)})^{s(d,v^{[d_1\backslash e]})}]^{w_2(d,d_1,v)}((v^{[d_1\backslash e]})^{[d\backslash e]} w_1(d, d_1, v))\} w_3(d, d_1, v), \tag{85}$$

where $w_3(d,d_1,v) = e/[\eta(s(d,v^{[d_1\backslash e]})w_2,s(d_1,v),b)\eta(s(d,v^{[d_1\backslash e]}),w_2,b^{s(d_1,v)})]$, $w_2 = w_2(d,d_1,v)$, $b = f((v^{[d_1\backslash e]})^{[d\backslash e]}w_1(d,d_1,v))$. Equations (68) and (61) imply that $(v\gamma)^{[a]} = v^{[a]}\gamma^{\tau}$ for each $v \in V$ and $\gamma \in \mathcal{C}(D)$, $a \in D$; consequently, $((vw_1)^{[d_1\backslash e]})^{[d\backslash e]} = (v^{[d_1\backslash e]})^{[d\backslash e]}w_1$, and hence, $vw_1 \in V$ for each $v \in V$, d and d_1 in D, $w_1 = w_1(d,d_1,v)$ by Equation (79), since $w_1 = \gamma^{\tau}$. Thus Equation (76) follows from Equations (78) and (85). □

Definition 4. *Suppose that the conditions of Remark 4 are satisfied and on the Cartesian product $C = D \times F$ (or $C^* = D \times F^*$) a binary operation is given by the following formula:*

$$(d_1,f_1)(d,f) = (d_1d,\xi((d_1^{\psi},f_1),(d^{\psi},f))f_1f^{\{d_1\}}), \tag{86}$$

where $\xi((d_1^{\psi},f_1),(d^{\psi},f))(v) = \xi((d_1^{\psi},f_1(v)),(d^{\psi},f(v)))$ for every d and d_1 in D, f and f_1 in F (or F^* respectively), and $v \in V$.

Theorem 5. *Let C, C^*, D, F, and F^* be the same as in Definition 4. Then, C and C^* are loops and there are natural embeddings $D \hookrightarrow C$, $F \hookrightarrow C$, $D \hookrightarrow C^*$, and $F^* \hookrightarrow C^*$ such that F (or F^*) is an almost normal subloop in C (or C^* respectively).*

Proof. The operation of Equation (86) is single-valued. Let $a = (d,f)$ and $b = (d_0,f_0)$, where d and d_0 are in D and where f and f_0 are in F (or F^*).

The equation $ay = b$ is equivalent to $dd_2 = d_0$ and

$$\xi((d^{\psi},f),(d_2^{\psi},f_2))ff_2^{\{d\}} = f_0,$$

where $d_2 \in D$, $f_2 \in F$ (or $f_2 \in F^*$ respectively), $y = (d_2,f_2)$, $\xi((d^{\psi},f),(d_2^{\psi},f_2))(v) = \xi((d^{\psi},f(v)),(d_2^{\psi},f_2(v)))$ for each $v \in V$. Therefore, $d_2 = d\backslash d_0$, $f_2^{\{d\}} = [\xi((d^{\psi},f),((d\backslash d_0)^{\psi},f_2))f]\backslash f_0$ by Equation (1) and Theorem 2. On the other hand, $f_2^{\{e\}} = f_2$ by Equation (74) and $f_2(v) = \{[(f_2^{\{d\}})^{\{d_3\}}]^{w_2}(vw_1)\}w_3$ by Equation (76), where $w_j = w_j(d,d_3,v)$, $j \in \{1,2,3\}$, $d_3 = d\backslash e$, and $dd_3 = e$ by Equation (14). Thus, using Equation (35), we get that

$$y = (d\backslash d_0,\{\{([\xi((d^{\psi},f),((d\backslash d_0)^{\psi},[(f\backslash f_0)^{\{d\backslash e\}}]^{w_2}w_3))f]\backslash f_0)^{\{d\backslash e\}}\}^{w_2}(vw_1)\}w_3)$$

belongs to C (or C^* respectively), giving Equation (1).

Then, we seek a solution $x \in C$ (or $x \in C^*$ respectively) of the equation $xa = b$. It is equivalent to two equations: $d_1d = d_0$ and

$$\xi((d_1^{\psi},f_1(v)),(d,f(v)))f_1(v)f^{\{d_1\}}(v) = f_0(v)$$

for each $v \in V$, where $d_1 \in D$, $f_1 \in F$ (or $f_1 \in F^*$ respectively), and $x = (d_1,f_1)$. Therefore, $d_1 = d_0/d$ and $f_1(v) = f_0(v)/[\xi(((d_0/d)^{\psi},f_1(v)),(d,f(v)))f^{\{d_0/d\}}(v)]$. Thus,

$$x = (d_0/d,f_0/[\xi(((d_0/d)^{\psi},f_1),(d,f))f^{\{d_0/d\}}])$$

belongs to C (or C^* respectively), giving Equation (2).

Moreover, $(e,e)(d,f) = (d,f)$ and $(d,f)(e,e) = (d,f)$ for each $d \in D$, $f \in F$ (or $f \in F^*$ respectively) by Equations (35) and (86). Therefore, the condition of Equation (3) is also satisfied. Thus, C and C^* are loops.

Evidently $D \ni d \mapsto (d,e)$ and $F \ni f \mapsto (e,f) \in C$ (or $F^* \ni f \mapsto (e,f) \in C^*$ respectively) provide embeddings of D and F (or D and F^* respectively) into C (or C^* respectively).

It remains to verify that F (or F^* respectively) is an almost normal subloop in C (or C^* respectively). Assume that $d_1 \in D$, $f_1 \in F$. Then,

$$(d_1, f_1)F = \{(d_1, \xi((d_1^{\psi}, f_1), (e, f))f_1 f^{\{d_1\}}) : f \in F\}$$

and

$$F(d_1, f_1) = \{(d_1, \xi((e, f), (d_1^{\psi}, f_1))f f_1) : f \in F\}.$$

Using the embedding $\mathcal{C}^V \hookrightarrow F$ and Equation (35), we infer that $(d_1, f_1)F = F(d_1, f_1)$, since $F^{\{d_1\}} = F$ by Equation (68), Lemma 5, and Equation (30). It can be verified similarly that F^* is the almost normal subloop in C^*. □

Definition 5. *The product Equation (86) in the loop C (or C^*) of Theorem 5 is called a smashed twisted wreath product of D and F (or a restricted smashed twisted wreath product of D and F^* respectively) with smashing factors ϕ, η, κ, and ξ, and it will be denoted by $C = D\Delta^{\phi,\eta,\kappa,\xi}F$ (or $C^* = D\Delta^{\phi,\eta,\kappa,\xi}F^*$ respectively). The loop C (or C^*) is also called a smashed splitting extension of F (or of F^* respectively) by D.*

Theorem 6. *Let the conditions of Remark 4 be satisfied and $\mathcal{C}_m(D) \subseteq C$, where C is as in Equation (28). Then, C and C^* supplied with the binary operation of Equation (86) are metagroups.*

Proof. In view of Theorem 5, C and C^* are loops. To each element b in B, there corresponds an element $\{b(v) : \forall v \in V,\ b(v) = b\}$ in F which can be denoted by b also. From the conditions of Equations (29)–(35), we deduce that

$$\gamma^a = \gamma \text{ and } f^{\gamma} = f \text{ for every } \gamma \in \mathcal{C} \text{ and } a \in A. \tag{87}$$

Hence, Equations (87) and (86) imply that $(\mathcal{C}(A), \mathcal{C}(F)) \subseteq \mathcal{C}(C)$. On the other hand, $w_1 = \gamma^{\tau}$ with $\gamma \in \mathcal{C}_m(D)$ and $w_2 = \gamma_3/(\gamma_1^{\psi}\gamma_2^{\psi}\gamma_5^{\psi})$ with $\gamma_1,...,\gamma_5$ in $\mathcal{C}_m(D)$ (see Equation (84)); hence, the condition $\mathcal{C}_m(D) \subset \mathcal{C}$ implies that Equation (76) simplifies to

$$f^{\{dd_1\}}(v) = (f^{\{d\}}(v))^{\{d_1\}} w_3(d, d_1, v) \tag{88}$$

for each $f \in F$, $v \in V$, and d and d_1 in D, since $\mathcal{C} \subseteq \mathcal{C}(A)$ by Equation (28). Next, we consider the following products:

$$I_1 = ((d_2, f_2)(d_1, f_1))(d, f) = ((d_2 d_1, \xi((d_2^{\psi}, f_2), (d_1^{\psi}, f_1))f_2 f_1^{\{d_2\}})(d, f) \tag{89}$$

and

$$I_2 = (d_2, f_2)((d_1, f_1)(d, f)) = (d_2, f_2)(d_1 d, \xi((d_1^{\psi}, f_1), (d^{\psi}, f))f_1 f^{\{d_1\}}). \tag{90}$$

Then, Equations (86), (90), and (33)–(35) imply that

$$I_2 = (d_2(d_1 d), \xi((d_1^{\psi}, f_1), (d^{\psi}, f))\xi((d_2^{\psi}, f_2), ((d_1 d)^{\psi}, f_1 f^{\{d_1\}}))\kappa(s(d_2, v), f_1(v^{[d_2\backslash e]}), f^{\{d_1\}}(v^{[d_2\backslash e]}))f_2(v)[f_1^{\{d_2\}}(v)(f^{\{d_1\}})^{\{d_2\}}(v)]. \tag{91}$$

From Equations (88), (89), (76), and (35), we infer that

$$I_1 = ((d_2 d_1)d, \xi((d_1^{\psi}, f_1), (d^{\psi}, f))\xi(((d_2 d_1)^{\psi}, f_2 f_1^{\{d_2\}}), (d^{\psi}, f))(f_2 f_1^{\{d_2\}})(f^{\{d_1\}})^{\{d_2\}} w_3, \tag{92}$$

where $w_3 = w_3(d_1, d_2, v)$. Therefore, from Equations (91) and (92), we infer that

$$I_1 = t_C((d_2, f_2), (d_1, f_1), (d, f))I_2, \tag{93}$$

where

$$t_C((d_2,f_2),(d_1,f_1),(d,f)) = t_D(d_2,d_1,d)t_B(f_2,f_1,f^{\{d_2d_1\}})\xi((d_1^{\psi},f_1),(d^{\psi},f))$$
$$\xi((d_2^{\psi},f_2),((d_1d)^{\psi},f_1f^{\{d_1\}}))\kappa(s(d_2,v),f_1(v^{[d_2\backslash e]}),f^{\{d_1\}}(v^{[d_2\backslash e]}))$$

$$/[\xi((d_2^{\psi},f_2),(d_1^{\psi},f_1))\xi(((d_2d_1)^{\psi},f_2f_1^{\{d_2\}}),(d^{\psi},f))w_3(d_1,d_2,v)]; \tag{94}$$

$$t_B(f_2,f_1,f)(v) = t_B(f_2(v),f_1(v),f(v)); \tag{95}$$

$$\xi((d_2^{\psi},f_2),(d_1^{\psi},f_1))(v) = \xi((d_2^{\psi},f_2(v)),(d_1^{\psi},f_1(v))) \tag{96}$$

for every f, f_1, f_2 in F, d, d_1, d_2 in D, and $v \in V$. Then from Equation (93), $\mathcal{C}(F) = (\mathcal{C}(B))^V$ (see Theorem 2) and Equation (28), it follows that the loops C and C* satisfy the condition of Equation (9), since $(\mathcal{C},\mathcal{C}^V) \subseteq \mathcal{C}(C)$. Thus, C and C* are metagroups. □

Remark 5. *Generally, if $A \neq \{e\}$ and $A \neq D$, B, ϕ, η, κ, and ξ are nontrivial, where A, B, and D are metagroups or particularly may be groups, then the loops C and C^* of Theorem 5 can be non-metagroups. If Equation (35) drops the conditions $\xi((e,e),(v,b)) = e$ and $\xi((v,b),(e,e)) = e$ for each $v \in V$ and $b \in B$, then the proofs of Theorems 3–5 demonstrate that C_1 and C_2 are strict quasi-groups and that C and C^* are quasi-groups.*

Definition 6. *Let P_1 and P_2 be two loops with centers $\mathcal{C}(P_1)$ and $\mathcal{C}(P_2)$. Let also*

$$\mu(a,b) = \nu(a,b)\mu(a)\mu(b) \tag{97}$$

for each a and b in P_1, where $\nu(a,b) \in \mathcal{C}(P_2)$. Then, μ will be called a metamorphism of P_1 into P_2. If in addition μ is surjective and bijective, then it will be called a metaisomorphism and it will be said that P_1 is metaisomorphic to P_2.

Theorem 7. *Suppose that A, B, and D are metagroups and that $A \subset D$, V_1, and V_2 are right transversals of A in D, $F_j = B^{V_j}$,*

$$P_j = D\Delta^{\phi,\eta,\kappa,\xi}F_j,\ P_j^* = D\Delta^{\phi,\eta,\kappa,\xi}F_j^*,\ j \in \{1,2\}.$$

Then, P_1 is metaisomorphic to P_2 and P_1^* to P_2^*.

Proof. By virtue of Theorem 5, P_j and P_j^* are loops, where $j \in \{1,2\}$, $\mathcal{C}^{V_j} \subset \mathcal{C}(P_j)$. From Equations (62) and (73), it follows that

$$s_j(\delta d,v) = s_j(d,v/\delta) = \delta^{\psi_j}s_j(d,v) \tag{98}$$

for each $d \in D$, $v \in V_j$, and $\delta \in \mathcal{C}(D)$, where s_j, $v^{[a]_j}$, d^{τ_j}, and d^{ψ_j} correspond to V_j, $j \in \{1,2\}$. Then, Equations (68) and (63) imply that

$$v^{[\delta/d]_j} = v^{[e/d]}\delta^{\tau_j} \tag{99}$$

for each $d \in D$, $v \in V_j$, and $\delta \in \mathcal{C}(D)$, $j \in \{1,2\}$. Therefore, from the identities of Equations (98), (99), and (84) and Lemma 2, we infer that

$$w_2(\delta d,\delta_1 d_1,\delta_2 v) = w_2(d,d_1,v) \tag{100}$$

for each of d and d_1 in D, δ, δ_1 and δ_2 in $\mathcal{C}(D)$, and $v \in V$.

For each of $f \in F_1$ and $v \in V_2$, we put

$$\mu f(v) = f^{e/v^{\psi_1}}(v^{\tau_1}). \tag{101}$$

From Lemma 5, it follows that $V_2^{\tau_1} = V_1$ and $v_1^{\tau_1} \neq v_2^{\tau_1}$ for each $v_1 \neq v_2$ in V_2, where $V_2^{\tau_1} = \{v^{\tau_1} : v \in V_2\}$. Then, Equations (87), (62), (100), and (101) and Lemma 1 imply that

$$f^{\{d\}\mu} = f^{\mu\{d\}} \tag{102}$$

for each $f \in F_1$, $d \in D$, where $f^{\mu\{d\}} = (\mu f)^{\{d\}}$, $f^{\{d\}\mu} = \mu(f^{\{d\}})$ (see also Equation (72)). From the identity of Equation (102) and the conditions of Equations (33) and (34), we infer that

$$\mu((d_1, f_1)(d, f))(v) = \kappa(e/v^{\psi_1}, f_1(v^{\tau_1}), f^{\{d_1\}}(v^{\tau_1}))(\mu(d_1, f_1))(\mu(d, f)) \tag{103}$$

for each of d and d_1 in D, f and f_1 in F_1, and $v \in V_2$, where $\mu(d, f) = (d, \mu f)$, $(d, f)(v) = (d, f(v))$. Hence,

$$\nu((d_1, f_1), (d, f))(v) = \kappa(e/v^{\psi_1}, f_1(v^{\tau_1}), f^{\{d_1\}}(v^{\tau_1})) \in \mathcal{C} \tag{104}$$

for each $v \in V_2$ (see also Equations (28) and (30)). Thus, P_1 is metaisomorphic to P_2 and P_1^* to P_2^*. □

Theorem 8. *Suppose that D is a nontrivial metagroup. Then, there exists a smashed splitting extension C^* of a nontrivial central metagroup H by D such that $[H, C^*]\mathcal{C}(H) = H$, where $[a, b] = (e/a)((e/b)(ab))$ for each a and b in C^*.*

Proof. Let d_0 be an arbitrary fixed element in $D - \mathcal{C}(D)$. Assume that A is a submetagroup in D such that A is generated by d_0 and a subgroup $\mathcal{C}_0$ contained in a center $\mathcal{C}(D)$ of D, $\mathcal{C}_m(D) \subseteq \mathcal{C}_0 \subseteq \mathcal{C}(D)$, where $\mathcal{C}_m(D)$ is a minimal subgroup in a center $\mathcal{C}(D)$ of D such that $t_D(a, b, c) \in \mathcal{C}_m(D)$ for each of a, b, and c in D. Therefore,

$$a^k a^n = p(k, n, a) a^{k+n} \tag{105}$$

for each $a \in A$, k, and n in $\mathcal{C} = \{0, -1, 1, -2, 2, ...\}$, where the following notation is used: $a^2 = aa$, $a^{n+1} = a^n a$ and $a^{-n} = e/a^n$, and $a^0 = e$ for each $n \in \mathbf{N}$ and $p(k, n, a) \in \mathcal{C}_m(A)$. Hence, in particular, A is a central metagroup. Then, $d_0\mathcal{C}_m(A)$ is a cyclic element in the quotient group $A/\mathcal{C}_m(A)$ (see Theorem 1). Then, we choose a central metagroup B generated by an element b_0 and a commutative group $\mathcal{C}_1$ such that $b_0 \notin \mathcal{C}_1$, $\mathcal{C}_m(D) \hookrightarrow \mathcal{C}_1$ and $\mathcal{C}(A) \hookrightarrow \mathcal{C}_1$ and the quotient group $B/\mathcal{C}_m(B)$ is of finite order $l > 1$. Then, let $\phi : A \to \mathcal{A}(B)$ satisfy the condition of Equation (30) and be such that

$$\phi(d_0) b_0 = b_0^2. \tag{106}$$

To satisfy the condition of Equation (106), a natural number l can be chosen as a divisor of $2^{|d_0\mathcal{C}_m(A)|} - 1$ if the order $|d_0\mathcal{C}_m(A)|$ of $d_0\mathcal{C}_m(A)$ in $A/\mathcal{C}_m(A)$ is positive; otherwise, l can be taken as any fixed odd number $l > 1$ if $A/\mathcal{C}_m(A)$ is infinite.

Then, we take a right transversal V of A in D so that A is represented in V by e. Let Ξ, η, κ, and ξ be chosen to satisfy the conditions of Equations (29)–(35), where $\mathcal{C}_m(B) \hookrightarrow \mathcal{C}$, $\mathcal{C}_m(A) \hookrightarrow \mathcal{C}$, $\mathcal{C}_0 \hookrightarrow \mathcal{C}$, and $\mathcal{C}_1 \hookrightarrow \mathcal{C}$. With these data, according to Theorem 6, C^* is a metagroup, since $\mathcal{C}_m(D) \hookrightarrow \mathcal{C}_1$ and $\mathcal{C}_m(D) \hookrightarrow \mathcal{C}_0$. That is, C^* is a smashed splitting extension of the central metagroup F^* by D.

Apparently, there exists $f_0 \in F^*$ such that $f_0(e) = b_0$, $f_0(v) = e$ for each $v \in V - \{e\}$. Therefore, $f_0^{\{v\}}(v) = b_0$ for each $v \in V$, since $s(v, v) = e$, $v^{[v\backslash e]} = [v(v\backslash e)]^\tau = e$.

Let $v_1 \neq v_2$ belong to V. Then, $(v_2(v_1\backslash e))^\tau = v_3 \in V$. Assume that $v_3 = e$. The latter is equivalent to $v_2(v_1\backslash e) = a \in A$. From Equation (13), it follows that $v_2 = a/(v_1\backslash e) = \gamma a v_1$, where $\gamma = t_D(v_1, v_1\backslash e, v_1)/t_D(av_1, v_1\backslash e, v_1)$ by Equation (11) and Lemma 2, since $e/(v_1\backslash e) = v_1$. Hence, $v_2 = v_2^\tau = (\gamma a v_1)^\tau = \gamma^\tau v_1$ by Equation (63), and consequently, $(v_2(v_1\backslash e))^\tau = \gamma^\tau = e$, contradicting the supposition $v_1 \neq v_2$. Thus, $v_3 \neq e$, and consequently, $f_0^{\{v_1\}}(v_2) = e^{s(v_1, v_2)} = e$ by Equation (31). This implies that $\{f_0^{\{v\}} : v \in V\}\mathcal{C}(F^*)$ generates F^*.

Evidently, $[v(d_0\backslash e)]^\tau \neq e$ for each $v \in V - \{e\}$, since $d_0\backslash e \in A$ and the following conditions $s \in D$, $sq \in A$, and $q \in A$ imply that $s \in A$ because A is the submetagroup in D. Note that

$e/d = (d \setminus e)/t_A(e/d, d, d\backslash e)$ for each $d \in A$ by Equation (11); consequently, $s(d, e) = dt_A(e/d, d, d\backslash e)$. On the other hand, $t_A(a, b, c) \in \mathcal{C}$ for each of a, b, and c in A and

$$f_0^\gamma = f_0 \; for \; each \; \gamma \in \mathcal{C} \tag{107}$$

by Equation (87); hence, $f_0^{\{d_0\}}(e) = \phi(d_0)b_0 = b_0^2$, and consequently,

$$f_0^{\{d_0\}} = f_0^2,$$

since

$$f_0^{\{d_0\}}(v) = e \; for \; each \; v \in V - \{e\}. \tag{108}$$

Therefore, we deduce using Equation (107) that

$$[(e, f_0), (e/d_0, e)] = (e, wf_0), \tag{109}$$

where

$$w = \xi((e, f_0), (e/d_0, e))\xi((d_0, e), (e/d_0, f_0))$$

$$\xi((e, e/f_0), (e, (f_0)^2))/t_{F^*}(e/f_0, f_0, f_0), \tag{110}$$

$$t_{F^*}(f, g, h)(v) = t_B(f(v), g(v), h(v)) \;\; for \; each \;\; v \in V, f, g \; and \; h \;\; in \;\; F^*. \tag{111}$$

Thus, $w = w(v) \in \mathcal{C}$ for each $v \in V$ and $f_0 \in [F^*, C^*]$, since $\mathcal{C}^V \cap F^* \subset \mathcal{C}(F^*)$. Hence, $F^* \subseteq [F^*, C^*]\mathcal{C}(F^*)$, since $F^* \hookrightarrow C^*$ and $\mathcal{C}(C^*) \cap F^* \subseteq \mathcal{C}(F^*)$. On the other hand, $\mathcal{C}_m(A) \hookrightarrow \mathcal{C}$, $\mathcal{C}_m(B) \hookrightarrow \mathcal{C}$, $\mathcal{C}_m(D) \hookrightarrow \mathcal{C}_j$, and $\mathcal{C}_j \hookrightarrow \mathcal{C}$ for each $j \in \{0, 1\}$. Therefore, Equations (107), (108), and (88) imply that $cF^* = F^*c$ and $c[F^*, C^*]\mathcal{C}(F^*) = [F^*, C^*]\mathcal{C}(F^*)c$ for each $c \in C^*$. Hence, $[F^*, C^*]\mathcal{C}(F^*) \subseteq F^*$. Taking $H = F^*$, we get the assertion of this theorem. □

Corollary 2. *Let the conditions of Theorem 8 be satisfied and D be generated by $\mathcal{C}_m(D)$ and at least two elements d_1, d_2,... such that $d_1 \neq e$ and $[d_2\backslash e, d_1\backslash e] = e$. Then, the smashed splitting extension C^* can be generated by $\mathcal{C}(F^*)$ and elements c_1, c_2,... such that $d_j\backslash e \in F^*c_j$ for each j.*

Proof. We take $d_0 = d_1$ in the proof of Theorem 8; thus, $c_1 = (d_1\backslash e, e)$, $c_2 = (d_2\backslash e, f_0)$, and $c_j = (d_j\backslash e, e)$ for each $j \geq 3$. Therefore Equations (66), (108), and (35) imply that

$$[c_2, c_1] = (e, pf_0), \text{ where}$$

$$p = \xi((d_2\backslash e, f_0), (d_1\backslash e, e))\xi((d_1, e), ((d_2\backslash e)(d_1 \setminus e), f_0))$$

$$\xi((e, e)/(d_2\backslash e, f_0), (d_2 \setminus e, (f_0)^2))/t_{F^*}(e/f_0, f_0, f_0), \tag{112}$$

since $[d_2\backslash e, d_1\backslash e] = e$ and $e/(d_2\backslash e) = d_2$. Thus, the submetagroup of C^* which is generated by $\mathcal{C}_m(D)$ and $\{c_j : j\}$ contains the metagroup D and (e, pf_0). Therefore, the following set $\{f^{\{d\}} : d \in D\}\mathcal{C}(F^*)$ generates the central metagroup F^*, since $V \subset D$ and $\{f^{\{v\}} : v \in V\}\mathcal{C}(F^*)$ generate F^*. Notice that $\mathcal{C}_m(D) \hookrightarrow \mathcal{C}(F^*)$. Hence, $\{c_j : j\}\mathcal{C}(F^*)$ generates C^*. □

Example 1. *Assume that A is a unital algebra over a commutative associative unital ring $\mathcal{F}$ supplied with a scalar involution $a \mapsto \bar{a}$ so that its norm N and trace T maps have values in $\mathcal{F}$ and fulfil conditions:*

$$a\bar{a} = N(a)1 \text{ with } N(a) \in \mathcal{F}, \tag{113}$$

$$a + \bar{a} = T(a)1 \text{ with } T(a) \in \mathcal{F}, \tag{114}$$

$$T(ab) = T(ba) \tag{115}$$

for each a and b in A.

We remind that, if a scalar $f \in F$ satisfies the condition $\forall a \in A\ fa = 0 \Rightarrow a = 0$, then such element f is called cancelable. For such a cancelable scalar f, the Cayley–Dickson doubling procedure induces a new algebra $C(A, f)$ over F such that

$$C(A, f) = A \oplus Al, \tag{116}$$

$$(a + bl)(c + dl) = (ac - f\bar{d}b) + (da + b\bar{c})l \tag{117}$$

and

$$\overline{(a + bl)} = \bar{a} - bl \tag{118}$$

for each a and b in A. Such an element l is called a doubling generator. From Equations (113)–(115), it follows that $\forall a \in A, \forall b \in A\ \ T(a) = T(a + bl)$ and $N(a + bl) = N(a) + fN(b)$. Apparently, the algebra A is embedded into $C(A, f)$ as $A \ni a \mapsto (a, 0)$, where $(a, b) = a + bl$. It is put by induction $A_n(f_{(n)}) = C(A_{n-1}, f_n)$, where $A_0 = A$, $f_1 = f$, $n = 1, 2, ...,$ and $f_{(n)} = (f_1, ..., f_n)$. Then, $A_n(f_{(n)})$ is a generalized Cayley–Dickson algebra, when F is not a field, or a Cayley–Dickson algebra, when F is a field.

There is an algebra $A_\infty(f) := \bigcup_{n=1}^\infty A_n(f_{(n)})$, where $f = (f_n : n \in \mathbf{N})$. In the case of $char(F) \neq 2$, let $Im(z) = z - T(z)/2$ be the imaginary part of a Cayley–Dickson number z and, hence, $N(a) := N_2(a, \bar{a})/2$, where $N_2(a, b) := T(a\bar{b})$.

If the doubling procedure starts from $A = F1 =: A_0$, then $A_1 = C(A, f_1)$ is a $*$-extension of F. If A_1 has a basis $\{1, u\}$ over F with the multiplication table $u^2 = u + w$, where $w \in F$ and $4w + 1 \neq 0$, with the involution $\bar{1} = 1$, $\bar{u} = 1 - u$, then A_2 is the generalized quaternion algebra and A_3 is the generalized octonion (Cayley–Dickson) algebra.

Particularly, for $F = \mathbf{R}$ and $f_n = 1$ for each n by $\mathcal{A}_r$ the real Cayley-Dickson algebra with generators $i_0, ..., i_{2^r-1}$ will be denoted such that $i_0 = 1$, $i_j^2 = -1$ for each $j \geq 1$, and $i_j i_k = -i_k i_j$ for each $j \neq k \geq 1$. Note that the Cayley–Dickson algebra $\mathcal{A}_r$ for each $r \geq 3$ is nonassociative, for example, $(i_1 i_2) i_4 = -i_1(i_2 i_4)$, etc. Moreover, for each $r \geq 4$, the Cayley–Dickson algebra $\mathcal{A}_r$ is nonalternative (see References [7,11,12]). Frequently, $\bar{a}$ is also denoted by a^* or $\tilde{a}$.

Then, $G_r = \{i_j, -i_j : j = 0, 1, ..., 2^r - 1\}$ is a finite metagroup for each $3 \leq r < \infty$. Equation (117) is an example of the smashed product.

Then, one can take a Cayley–Dickson algebra A_n over a commutative associative unital ring $\mathcal{R}$ of characteristic different from two such that $A_0 = \mathcal{R}$, $n \geq 2$. There are basic generators $i_0, i_1, ..., i_{2^n-1}$, where $i_0 = 1$. Choose $\mathbf{\Psi}$ as a multiplicative subgroup contained in the ring $\mathcal{R}$ such that $f_j \in \mathbf{\Psi}$ for each $j = 0, ..., n$. Put $G_n = \{i_0, i_1, ..., i_{2^n-1}\} \times \mathbf{\Psi}$. Then, G_n is a central metagroup because, in this case, $\mathbf{\Psi}$ is commutative.

Example 2. *More generally, suppose that H is a group such that $\mathbf{\Psi} \subset H$, with relations $hi_k = i_k h$ and $(hg)i_k = h(gi_k)$ for each $k = 0, 1, ..., 2^n - 1$ and each h and g in H. Then, $G_n = \{i_0, i_1, ..., i_{2^n-1}\} \times H$ is also a metagroup. If the group H is noncommutative, then the latter metagroup can be noncentral (see the condition of Equation (10) in Definition 1). Utilizing the notation of Example 1, we get that the Cayley–Dickson algebra $\mathcal{A}_\infty$ over the real field $\mathbf{R}$ with $f_n = 1$ for each n provides a pattern of a metagroup $G_\infty = \{i_j, -i_j : 0 \leq j \in \mathbf{Z}\}$, where $\mathbf{Z}$ denotes the ring of integers.*

Example 3. *Certainly, in general, metagroups need not be central. On the other hand, if a metagroup is associative, then it is a group [1]. Apparently, each group is a metagroup also. For a group G, its associativity evidently means that $t_G(a, b, c) = e$ [1].*

From the given metagroups, new metagroups can be constructed using their direct, semidirect products, smashed products, and smashed twisted wreath products. Therefore, there are abundant families of noncentral metagroups and also of central metagroups different from groups.

Equations (39), (46), (47), (85), (86), and (94)–(96) provide examples of metagroups with complicated nonassociative noncommutative structures. The presented above theorems also permit to construct different examples of nonassociative quasi-groups and loops.

5. Conclusions

The results of this article can be used for further studies of metagroups, quasi-groups, loops, and noncommutative manifolds related with them. Besides applications of metagroups, loops, and quasi-groups outlined in the introduction, it is interesting to mention possible applications in mathematical coding theory and classification of information flows and their technological implementations [28–30] because, frequently, codes are based on binary systems. Moreover, twisted products are used for creating complicated codes [22]. In view of this, to study creating more complicated codes with the help of smashed twisted products of metagroups, Equations (86) and (94)–(96) provide additional options in the nonassociative case in comparison with the associative case.

Wreath products of groups are used for studies of varieties [24], so it will be interesting to investigate noncommutative varieties using metagroups. Then, twisted products are utilized for investigations of Lie groups and semi-Riemann manifolds [23,25]. Therefore, we will study their nonassociative metagroup analogs that can be used in noncommutative geometry and quantum field theory [16,31–35] because Lie groups and manifolds are actively used in these areas.

Funding: This research received no external funding.

Conflicts of Interest: The author declares no conflict of interest.

References

1. Bruck, R.H. *A Survey of Binary Systems*; Springer: Berlin, Germany, 1971.
2. Kakkar, V. Boolean loops with compact left inner mapping groups are profinite. *Topol. Appl.* **2018**, *244*, 51–54. [CrossRef]
3. Razmyslov, Y.P. *Identities of Algebras and their Representations*; Series Modern Algebra; Nauka: Moscow, Soviet Union, 1989; Volume 14.
4. Vojtěchoivský, P. Bol loops and Bruck loops of order *pq* up to isotopism. *Finite Fields Appl.* **2018**, *52*, 1–9.
5. Pickert, G. *Projektive Ebenen*; Springer: Berlin, Germany, 1955.
6. Pickert, G. Doppelebenen und loops. *J. Geom.* **1991**, *41*, 133–144. [CrossRef]
7. Baez, J.C. The octonions. *Bull. Am. Math. Soc.* **2002**, *39*, 145–205. [CrossRef]
8. Bogolubov, N.N.; Logunov, A.A.; Oksak, A.I.; Todorov, I.T. *General Principles of Quantum Field Theory*; Nauka: Moscow, Russia, 1987.
9. Bourbaki, N. *Algebra*; Springer: Berlin, Germany, 1989.
10. Castro-Alvaredo, O.A.; Doyon, B.; Fioravanti, D. Conical twist fields and null polygonal Wilson loops. *Nuclear Phys.* **2018**, *B931*, 146–178. [CrossRef]
11. Dickson, L.E. *The Collected Mathematical Papers*; Chelsea Publishing Co.: New York, NY, USA, 1975; Volumes 1–5.
12. Kantor, I.L.; Solodovnikov, A.S. *Hypercomplex Numbers*; Springer: Berlin, Germany, 1989.
13. Schafer, R.D. *An Introduction to Nonassociative Algebras*; Academic Press: New York, NY, USA, 1966.
14. Frenod, E.; Ludkowski, S.V. Integral operator approach over octonions to solution of nonlinear PDE. *Far East J. Math. Sci. (FJMS)* **2018**, *103*, 831–876. [CrossRef]
15. Gürlebeck, K.; Sprössig, W. *Quaternionic and Clifford Calculus for Physicists and Engineers*; John Wiley and Sons: Chichester, UK, 1997.
16. Gürsey, F.; Tze, C.-H. *On the Role of Division, Jordan and Related Algebras in Particle Physics*; World Scientific Publishing Co.: Singapore, 1996.
17. Ludkowski, S.V. Integration of vector Sobolev type PDE over octonions. *Complex Var. Elliptic Equ.* **2016**, *61*, 1014–1035.
18. Ludkovsky, S.V.; Sprössig, W. Spectral theory of super-differential operators of quaternion and octonion variables. *Adv. Appl. Clifford Algebras* **2011**, *21*, 165–191.

19. Ludkovsky, S.V. Integration of vector hydrodynamical partial differential equations over octonions. *Complex Var. Elliptic Equ.* **2013**, *58*, 579–609. [CrossRef]
20. Ludkowski, S.V. Automorphisms and derivations of nonassociative C^* algebras. *Linear Multilinear Algebra* **2019**, *67*, 1531–1538. [CrossRef]
21. Ludkowski, S.V. Cohomology theory of nonassociative algebras. *Axioms* **2019**, *8*, 78. [CrossRef]
22. Betten, A. Twisted tensor product codes. *Des. Codes Cryptogr.* **2008**, *47*, 191–219. [CrossRef]
23. Fernández-López, M.; García-Río, E.; Kupeli, D.N.; Ünal, B. A curvature condition for a twisted product to be a warped product. *Manuscripta Math.* **2001**, *106*, 213–217. [CrossRef]
24. Mikaelian, V.H. The criterion of Shmel'kin and varieties generated by wreath products of finite groups. *Algebra Logic* **2017**, *56*, 108–115. [CrossRef]
25. Rudkovski, M.A. Twisted products of Lie groups. *Sib. Math. J.* **1997**, *38*, 969–977. [CrossRef]
26. Othman, M.I.A.; Marin, M. Effect of thermal loading due to laser pulse on thermoelastic porous medium under G-N theory. *Results Phys.* **2017**, *7*, 3863–3872. [CrossRef]
27. Kunen, K. *Set Theory*; North-Holland Publishing Co.: Amsterdam, The Netherlands, 1980.
28. Blahut, R.E. *Algebraic Codes for Data Transmission*; Cambridge University Press: Cambridge, UK, 2003.
29. Shum, K.P.; Ren, X.; Wang, Y. Semigroups on semilattice and the constructions of generalized cryptogroups. *Southeast Asian Bull. Math.* **2014**, *38*, 719–730.
30. Sigov, A.S.; Andrianova, E.G.; Zhukov, D.O.; Zykov, S.V.; Tarasov, I.E. Quantum informatics: Overview of the main achievements. *Russ. Technol. J.* **2019**, *7*, 5–37. [CrossRef]
31. Gilbert, J.E.; Murray, M.A.M. *Clifford Algebras and Dirac Operators in Harmonic Analysis*; Cambridge Studies in Advanced Mathematics Book 26; Cambridge University Press: Cambridge, UK, 1991.
32. Girard, P.R. *Quaternions, Clifford Algebras and Relativistic Physics*; Birkhäuser: Basel, Switzerland, 2007.
33. Ludkowski, S.V. Manifolds over Cayley-Dickson algebras and their immersions. *Mathematics* **2013**, arXiv:1204.1545.
34. Ludkovsky, S.V. Normal families of functions and groups of pseudoconformal diffeomorphisms of quaternion and octonion variables. *J. Math. Sci. N. Y. (Springer)* **2008**, *150*, 2224–2287. [CrossRef]
35. Ludkovsky, S.V. Functions of several Cayley-Dickson variables and manifolds over them. *J. Math. Sci.* **2007**, *141*, 1299–1330. [CrossRef]

axioms

MDPI

Article

Unification Theories: Examples and Applications

Florin F. Nichita

Simion Stoilow Institute of Mathematics of the Romanian Academy 21 Calea Grivitei Street, 010702 Bucharest, Romania; Florin.Nichita@imar.ro; Tel.: +40-0-213-196-506; Fax: +40-0-213-196-505

Received: 24 October 2018; Accepted: 13 November 2018; Published: 16 November 2018

Abstract: We consider several unification problems in mathematics. We refer to transcendental numbers. Furthermore, we present some ways to unify the main non-associative algebras (Lie algebras and Jordan algebras) and associative algebras.

Keywords: transcendental numbers; Euler formula; Yang–Baxter equation; Jordan algebras; Lie algebras; associative algebras; coalgebras

MSC: 17C05; 17C50; 16T15; 16T25; 17B01; 03B05; 51A05; 68-04

---

1. Introduction

Andre Weil explains (see [1]) that the areas of modern mathematics need to be unified in a simple and general theory. In this way, we will see more clearly which are the main problems in mathematics (as opposed to the less important themes and results).

In physics, unification theories have contributed to a better understanding of certain phenomena. The unified approaches suggested in this communication refer to the theory of functions and algebraic structures, two of the pillars of modern physical tools.

The purpose of this paper is three-fold: (i) we present new results on transcendental numbers; (ii) we obtain new results in the theory of the unification of non-associative structures; (iii) we re-initiate a debate on unification(s) in mathematics.

In the next section, we will give several examples of unification problems. This section is related to the paper [2] on transcendental numbers. Some knowledge of Hopf algebra theory is needed in order to understand some results from this section.

The third section presents structures that unify (non-)associative structures. The main non-associative structures are Lie algebras and Jordan algebras. Arguably less studied, Jordan algebras have applications in physics, differential geometry, ring geometries, quantum groups, analysis, biology, etc. (see [3]). There are several ways to unify Lie algebras, Jordan algebras and associative algebras. We will also refer to cases when the unification of (non-)associative structures could be realized just in the conclusions of theorems.

This paper is related to a talk and a poster presented at "Noncommutative and non-associative structures, braces and applications", Malta, 11–15 March 2018, and to a private communication made at the "Workshop on Non-associative Algebras and Applications", Lancaster University, UK, July 2018.

All tensor products will be defined over the field k.

2. Examples of Unification Problems

In this section, we will give several examples of unification problems related to transcendental numbers.

The following two identities with transcendental numbers:

$$\int_{-\infty}^{+\infty} e^{-x^2} dx = \sqrt{\pi}\,, \tag{1}$$

$$\int_{-\infty}^{+\infty} e^{-ix^2} dx = \sqrt{\frac{\pi}{2}}(1-i)\,, \tag{2}$$

were unified and proven (by contour integration) in [4].

The next formulas (from [2]) can be also unified:

$$e^{\pi i} + 1 = 0\,, \tag{3}$$

$$|e^i - \pi| > e \tag{4}$$

and:

$$|\, e^{1-z} + e^{\bar{z}}\,| > \pi \qquad \forall z \in \mathbb{C}\,. \tag{5}$$

Remark 1. *Let us consider the two-variable complex function* $f : \mathbb{C} \times \mathbb{C} \to \mathbb{R}$, $f(z,w) = |e^z + e^w|$, *which gives the length of the sum of the vectors* e^z *and* e^w. *The Formulas (3)–(5) can be unified using the function* $f(z,w)$:

$$f(\pi i,\, 0) = 0,\;\; f(1-z, \bar{z}) > \pi \;\; \forall z \in \mathbb{C}, \quad f(i,\, \pi i + \ln \pi) > e.$$

Remark 2. *The function* $f(z,w)$ *can be expressed in another form using the formula* $|e^{x+i\alpha} + e^{y+i\beta}| = \rho\sqrt{1+\sin(2\theta)\cos(\alpha-\beta)}$, *where* $\rho = \sqrt{e^{2x}+e^{2y}}$ *and* $\theta = \cos^{-1}\left(\frac{e^x}{\rho}\right)$. *The relations from Remark 1 can be reinterpreted using this formula. For example, in the first formula:* $\rho = \sqrt{2}$, $\theta = \frac{\pi}{4}$, $\alpha = \pi$, $\beta = 0$; *so,* $f(\pi i,\, 0) = \sqrt{2}\sqrt{1+\sin(\frac{\pi}{2})\cos(\pi)} = 0$.

Remark 3. *While properties about the image of the function* $f(z,w)$ *unify the Formulas (3)–(5), the formula* $e^{x+iy} = e^x(\cos y + i \sin y)$ *could be considered a common part (or an essential part) of all of them.*

This latest formula can be related to a certain subcoalgebra of the trigonometric coalgebra (see [5]). Indeed, the properties of the trigonometric functions cos *and* sin *lead to the trigonometric coalgebra, given by the maps:* $\Delta(c) = c \otimes c - s \otimes s$, $\Delta(s) = s \otimes c + c \otimes s$, $\varepsilon(c) = 1$, $\varepsilon(s) = 0$. *Euler's relation leads to the subcoalgebra generated by* $c + is$.

The dual case states that $1 + ix$ *generates an ideal in the* $\mathbb{C}$ *algebra* $\frac{\mathbb{C}[X]}{X^2+1} = \mathbb{C}[x]$, *where* $x^2 = -1$. *In other words, Euler's relation implies that* $\forall a, b \in \mathbb{C}$, *there exists* $c \in \mathbb{C}$ *such that* $(a+bx)(1+ix) = c(1+ix)$ *(this can be checked directly).*

According to [5], the role of such objects in number theory is unexplored at the moment.

Remark 4. *The properties of the hyperbolic functions* cosh *and* sinh *lead to the following coalgebra, given by the maps:* $\Delta(c) = c \otimes c + s \otimes s$, $\Delta(s) = s \otimes c + c \otimes s$, $\varepsilon(c) = 1$, $\varepsilon(s) = 0$. *There exists a subcoalgebra generated by* $c + s$, *which can be related to Theorem 1 of [2], leading to some kind of "Euler formula" for hyperbolic functions.*

Remark 5. *The coalgebras from Remarks 3 and 4 can be unified as follows. For* $a \in k$, *we consider the coalgebra generated by* c *and* s, $\Delta(c) = c \otimes c + a^2 s \otimes s$, $\Delta(s) = s \otimes c + c \otimes s$, $\varepsilon(c) = 1$, $\varepsilon(s) = 0$. *There exists a subcoalgebra generated by* $c + as$.

Remark 6. *The following inequalities hold (see also [2]):* $\pi > |e^i - \pi| > e$.

The last inequality was proven in [2]: $|e^i - \pi| > e \iff (\pi+e)(\pi-e) + 1 > 2\pi\cos 1$. *Now,* $\pi > 3.141$, $e < 2.719$; *so,* $(\pi+e)(\pi-e)+1 > 3.47292$ *and* $\cos 1 < 1 - \frac{1}{2} + \frac{1}{4!} = \frac{13}{24}$, $2\pi\cos 1 < 3.142 \times \frac{13}{12} = 3.4038(3)$.

The inequality $\pi > |e^i - \pi|$ *is equivalent to* $2\pi\cos 1 > 1$, *which follows from* $\cos 1 > \cos 60^\circ = \frac{1}{2}$.

Remark 7. *The next formula generalizes the Basel problem* $\lim_{n\to\infty}\sum_{k=1}^{n}\frac{1}{k^2}=\frac{\pi^2}{6}$:

$$\sum_{1}^{n}\frac{1}{k^2}<\frac{2}{3}\left(\frac{n+1}{n}\right)^n \quad \forall n\in\mathbb{N}^*. \tag{6}$$

It could be an interesting problem to prove a similar formula for non-associative algebras. Likewise, one could try to generalize it for q-shifted factorials (see, for example, [6]).

3. The Unification of Non-Associative Structures

3.1. UJLA Structures

The UJLA (Unification of Jordan, Lie (and) Associative (algebras)) structures could be seen as structures that comprise the information encapsulated in associative algebras, Lie algebras and Jordan algebras. Thus, the category of the UJLA structures can be seen as a category that "includes" the categories of Jordan, Lie and associative algebras. A motivation for this unification is related to the study of bundles over Grassmannian manifolds.

Definition 1. *For a k-space V, let* $\eta : V\otimes V\to V,\ a\otimes b\mapsto ab$, *be a linear map such that:*

$$(ab)c+(bc)a+(ca)b=a(bc)+b(ca)+c(ab), \tag{7}$$

$$(a^2b)a=a^2(ba),\quad (ab)a^2=a(ba^2),\quad (ba^2)a=(ba)a^2,\quad a^2(ab)=a(a^2b), \tag{8}$$

$\forall\, a,b,c\in V$. *Then,* (V,η) *is called a UJLA structure.*

Remark 8. *If* $(A,\ \theta)$, *where* $\theta : A\otimes A\to A,\ \theta(a\otimes b)=ab$, *is a (non-unital) associative algebra, then we define a UJLA structure* $(A,\ \theta')$, *where* $\theta'(a\otimes b)=\alpha ab+\beta ba$, *for some* $\alpha,\ \beta\in k$. *For* $\alpha=\beta=\frac{1}{2}$, $(A,\ \theta')$ *is a Jordan algebra, and for* $\alpha=1=-\beta$, $(A,\ \theta')$ *is a Lie algebra.*

Theorem 1. *(Nichita [7]) Let* (V,η) *be a UJLA structure. Then,* $(V,\eta'),\ \eta'(a\otimes b)=[a,b]=ab-ba$ *is a Lie algebra.*

Theorem 2. *(Nichita [7]) Let* (V,η) *be a UJLA structure. Then,* $(V,\eta'),\ \eta'(a\otimes b)=a\circ b=\frac{1}{2}(ab+ba)$ *is a Jordan algebra.*

Remark 9. *The structures from the two above theorems are related by the relation:*

$$[a,b\circ c]+[b,c\circ a]+[c,a\circ b]=0.$$

Remark 10. *The classification of UJLA structures is an open problem.*

Remark 11. *If the characteristic of k is two, then a Lie algebra is also a Jordan algebra.*

Proof. Because the characteristic of k is two, the Lie algebra L is also commutative.

It follows easily that $[[x,x],x]=0\ \forall x\in L$.

Now, in the Jacobi identity, we take $z=x^2$: $[[x,y],x^2]+[[y,x^2],x]+[[x^2,x],y]=0$.

From the above observations, it follows that $[[x,y],x^2]=[x,[y,x^2]]$. Therefore, L is also a Jordan algebra. □

3.2. Yang–Baxter Equations

The authors of [8] argued that the Yang–Baxter equation leads to another unification of (non-)associative structures.

For V a k-space, we denote by $\tau : V \otimes V \to V \otimes V$ the twist map defined by $\tau(v \otimes w) = w \otimes v$ and by $I : V \to V$ the identity map of the space V; for $R : V \otimes V \to V \otimes V$ a k-linear map, let $R^{12} = R \otimes I$, $R^{23} = I \otimes R$, $R^{13} = (I \otimes \tau)(R \otimes I)(I \otimes \tau)$.

Definition 2. *A The Yang–Baxter operator is an invertible k-linear map, $R : V \otimes V \to V \otimes V$, which satisfies the braid condition (sometimes called the Yang–Baxter equation):*

$$R^{12} \circ R^{23} \circ R^{12} = R^{23} \circ R^{12} \circ R^{23}. \tag{9}$$

If R satisfies (9), then both $R \circ \tau$ and $\tau \circ R$ satisfy the quantum Yang–Baxter equation (QYBE):

$$R^{12} \circ R^{13} \circ R^{23} = R^{23} \circ R^{13} \circ R^{12}. \tag{10}$$

Therefore, Equations (9) and (10) are equivalent.

Let A be a (unitary) associative k-algebra, and $\alpha, \beta, \gamma \in k$; the authors of [9] defined the k-linear map $R^A_{\alpha,\beta,\gamma} : A \otimes A \to A \otimes A$,

$$a \otimes b \mapsto \alpha ab \otimes 1 + \beta 1 \otimes ab - \gamma a \otimes b \tag{11}$$

which is a Yang–Baxter operator if and only if one of the following cases holds:

(i) $\alpha = \gamma \neq 0, \ \beta \neq 0$;
(ii) $\beta = \gamma \neq 0, \ \alpha \neq 0$;
(iii) $\alpha = \beta = 0, \ \gamma \neq 0$.

An interesting property of (11), can be visualized in knot theory, where the link invariant associated with $R^A_{\alpha,\beta,\gamma}$ is the Alexander polynomial.

For $(L, [,])$ a Lie algebra over k, $z \in Z(L) = \{z \in L : [z, x] = 0 \ \forall\, x \in L\}$, and $\alpha \in k$, the authors of the papers [10,11] defined the following Yang–Baxter operator: $\phi^L_\alpha : L \otimes L \longrightarrow L \otimes L$,

$$x \otimes y \mapsto \alpha[x, y] \otimes z + y \otimes x\,. \tag{12}$$

Remark 12. *The Formulas (11) and (12) lead to the unification of associative algebras and Lie algebras in the framework of Yang–Baxter structures. At this moment, we do not have a satisfactory answer to the question how Jordan algebras fit in this framework (several partial answers were given).*

3.3. Unification of the Conclusions of Theorems

Sometimes, it is not easy to find structures that unify theorems for (non-)associative structures, but we could unify just the conclusions of theorems, as we will see in the next theorems.

Theorem 3. *If A is a Jordan algebra, a Lie algebra or an associative algebra and if $a, b \in A$, then:*

$$D : A \to A, \ \ D(x) = a(bx) + b(ax) + (ax)b - a(xb) - (xb)a - (xa)b$$

is a derivation.

Proof. We consider three cases. If A is a Jordan algebra, then $D(x) = a(bx) + b(ax) + (ax)b - a(xb) - (xb)a - (xa)b = a(bx) - (xa)b = a(bx) - b(ax)$. According to [12], D is a derivation.

If A is a Lie algebra, then $D(x) = a(bx) + b(ax) + (ax)b - a(xb) - (xb)a - (xa)b = a(bx) - b(ax) = a(bx) + b(xa) = (ab)x$. Therefore, D is a derivation.

If A is an associative algebra, then $D(x) = a(bx) + b(ax) + (ax)b - a(xb) - (xb)a - (xa)b = (ab + ba)x - x(ab + ba)$. Therefore, D is a derivation. □

Theorem 4. *If A is a Jordan algebra, a Lie algebra or an associative algebra and if $a, b \in A$, then $D : A \to A$, $D(x) = a(bx) - (xa)b$ is a derivation.*

Proof. The proof follows the same argumentation as the previous one, for which reason it is omitted. □

Remark 13. *We presented unifications of formulas, structures, categories and theorems. It is currently an open problem whether these types can be formulated in a unified form.*

Funding: This research received no external funding.

Acknowledgments: We would like to thank the referees for their high quality comments and the Editors for their help.

Conflicts of Interest: The author declares no conflict of interest.

References

1. Weil, A. A 1940 Letter of Andre Weil on Analogy in Mathematics (Trans. by Martin H. Krieger). *Notices AMS* **2005**, *52*, 341. (Trans. in Romanian by Florin Caragiu, "Cunoasterea Stiintifica in Orizontul Experierii Tainei (II), pg. 98).
2. Marcus, S.; Nichita, F.F. On Transcendental Numbers: New Results and a Little History. *Axioms* **2018**, *7*, 15. [CrossRef]
3. Iordanescu, R. *Jordan Structures in Geometry and Physics with an Appendix on Jordan Structures in Analysis*; Romanian Academy Press: Bucharest, Romania, 2003.
4. Desbrow, D. On Evaluating $\int_{-\infty}^{+\infty} e^{ax(x-2b)}dx$ by Contour Integration Round a Parallelogram. *Am. Math. Mon.* **1998**, *105*, 726–731.
5. Majid, S. *A Quantum Groups Primer*; Cambridge University Press: Cambridge, UK, 2002.
6. Ernst, T. Convergence Aspects for Generalizations of q-Hypergeometric Functions. *Axioms* **2015**, *4*, 134–155. [CrossRef]
7. Nichita, F.F. On Jordan algebras and unification theories. *Roman. J. Pure Appl. Math.* **2016**, *4*, 305–316.
8. Iordanescu, R.; Nichita, F.F.; Nichita, I.M. The Yang–Baxter Equation, (Quantum) Computers and Unifying Theories. *Axioms* **2014**, *3*, 360–368. [CrossRef]
9. Dăscălescu, S; Nichita, F.F. Yang–Baxter operators arising from (co)algebra structures. *Commun. Algebra* **1999**, *27*, 5833–5845. [CrossRef]
10. Majid, S. Solutions of the Yang–Baxter equation from braided-Lie algebras and braided groups. *J. Knot Theory Its Ramif.* **1995**, *4*, 673–697. [CrossRef]
11. Nichita, F.F.; Popovici, B.P. Yang–Baxter operators from (G, θ)-Lie algebras. *Roman. Rep. Phys.* **2011**, *63*, 641–650.
12. Todorov, I.; Dubois-Violette, M. Deducing the symmetry of the standard model from the automorphism and structure groups of the exceptional Jordan algebra. *arXiv* **2018**, arXiv:1806.09450.

axioms

MDPI

Article

Unification Theories: New Results and Examples

Florin F. Nichita

Simion Stoilow Institute of Mathematics of the Romanian Academy 21 Calea Grivitei Street, 010702 Bucharest, Romania; florin.nichita@imar.ro; Tel.: +40-0-21-319-6506; Fax: +40-0-21-319-6505

Received: 3 May 2019; Accepted: 17 May 2019; Published: 18 May 2019

Abstract: This paper is a continuation of a previous article that appeared in AXIOMS in 2018. A Euler's formula for hyperbolic functions is considered a consequence of a unifying point of view. Then, the unification of Jordan, Lie, and associative algebras is revisited. We also explain that derivations and co-derivations can be unified. Finally, we consider a "modified" Yang–Baxter type equation, which unifies several problems in mathematics.

Keywords: Euler's formula; hyperbolic functions; Yang–Baxter equation; Jordan algebras; Lie algebras; associative algebras; UJLA structures; (co)derivation

MSC: 17C05; 17C50; 16T15; 16T25; 17B01; 17B40; 15A18; 11J81

---

1. Introduction

Voted the most famous formula by undergraduate students, the Euler's identity states that $e^{\pi i} + 1 = 0$. This is a particular case of the Euler's–De Moivre formula:

$$\cos x + i \sin x = e^{ix} \quad \forall x \in \mathbb{R}, \tag{1}$$

and, for hyperbolic functions, we have an analogous formula:

$$\cosh x + J \sinh x = e^{xJ} \quad \forall x \in \mathbb{C}, \tag{2}$$

where we consider the matrices

$$J = \begin{pmatrix} 0 & 0 & 0 & 1 \\ 0 & 0 & 1 & 0 \\ 0 & 1 & 0 & 0 \\ 1 & 0 & 0 & 0 \end{pmatrix} \tag{3}$$

$$I = \begin{pmatrix} 1 & 0 & 0 & 0 \\ 0 & 1 & 0 & 0 \\ 0 & 0 & 1 & 0 \\ 0 & 0 & 0 & 1 \end{pmatrix} \tag{4}$$

$$I' = \begin{pmatrix} 1 & 0 \\ 0 & 1 \end{pmatrix} . \tag{5}$$

In fact, $R(x) = \cosh(x)I + \sinh(x)J = \cosh x + J \sinh x = e^{xJ}$ also satisfies the equation

$$(R \otimes I')(x) \circ (I' \otimes R)(x+y) \circ (R \otimes I')(y) = (I' \otimes R)(y) \circ (R \otimes I')(x+y) \circ (I' \otimes R)(x) \tag{6}$$

called the colored Yang–Baxter equation. This fact follows easily from $J^{12} \circ J^{23} = J^{23} \circ J^{12}$ and $xJ^{12} + (x+y)\, J^{23} + yJ^{12} = yJ^{23} + (x+y)\, J^{12} + xJ^{23}$, and it shows that the formulas (1) and (2) are related.

While we do not know a remarkable identity related to (2), let us recall an interesting inequality from a previous paper: $|e^i - \pi| > e$. There is an open problem to find the matrix version of this inequality.

The above analysis is a consequence of a unifying point of view from previous papers ([1,2]).

In the remainder of this paper, we first consider the unification of the Jordan, Lie, and associative algebras. In Section 3, we explain that derivations and co-derivations can be unified. We suggest applications in differential geometry. Finally, we consider a "modified" Yang–Baxter equation which unifies the problem of the three matrices, generalized eigenvalue problems, and the Yang–Baxter matrix equation. There are several versions of the Yang–Baxter equation (see, for example, [3,4]) presented throughout this paper.

We work over the field k, and the tensor products are defined over k.

2. Weak Ujla Structures, Dual Structures, Unification

Definition 1. *(Ref. [5]) Given a vector space V, with a linear map* $\eta : V \otimes V \to V,\ \eta(a \otimes b) = ab$, *the couple* $(V,\ \eta)$ *is called a "weak UJLA structure" if the product* $ab = \eta(a \otimes b)$ *satisfies the identity*

$$(ab)c + (bc)a + (ca)b \;=\; a(bc) + b(ca) + c(ab) \quad \forall\, a, b, c \in V. \tag{7}$$

Definition 2. *Given a vector space V, with a linear map* $\Delta : V \to V \otimes V$, *the couple* $(V,\ \Delta)$ *is called a "weak co-UJLA structure" if this co-product satisfies the identity*

$$(Id + S + S^2) \circ (\Delta \otimes I) \circ \Delta = (Id + S + S^2) \circ (I \otimes \Delta) \circ \Delta \tag{8}$$

where $S : V \otimes V \otimes V \to V \otimes V \otimes V,\ a \otimes b \otimes c \mapsto b \otimes c \otimes a$, $I : V \to V,\ a \mapsto a$ *and* $Id : V \otimes V \otimes V \to V \otimes V \otimes V,\ a \otimes b \otimes c \mapsto a \otimes b \otimes c$.

Definition 3. *Given a vector space V, with a linear map* $\phi : V \otimes V \to V \otimes V$, *the couple* $(V,\ \phi)$ *is called a "weak (co)UJLA structure" if the map* ϕ *satisfies the identity*

$$(Id + S + S^2) \circ \phi^{12} \circ \phi^{23} \circ \phi^{12} \circ (Id + S + S^2) = (Id + S + S^2) \circ \phi^{23} \circ \phi^{12} \circ \phi^{23} \circ (Id + S + S^2) \tag{9}$$

where $\phi^{12} = \phi \otimes I,\ \phi^{23} = I \otimes \phi$, $Id : V \otimes V \otimes V \to V \otimes V \otimes V,\ a \otimes b \otimes c \mapsto a \otimes b \otimes c$ *and* $I : V \to V,\ a \mapsto a$.

Theorem 1. *Let* (V, η) *be a weak UJLA structure with the unity* $1 \in V$. *Let* $\phi : V \otimes V \to V \otimes V,\ a \otimes b \mapsto ab \otimes 1$. *Then,* (V, ϕ) *is a "weak (co)UJLA structure".*

Proof. $(Id + S + S^2) \circ \phi^{23} \circ \phi^{12} \circ \phi^{23} \circ (Id + S + S^2)(a \otimes b \otimes c) = (Id + S + S^2) \circ \phi^{23} \circ \phi^{12} \circ \phi^{23}(a \otimes b \otimes c + b \otimes c \otimes a + c \otimes a \otimes b) = (Id + S + S^2) \circ \phi^{23} \circ \phi^{12}(a \otimes bc \otimes 1 + b \otimes ca \otimes 1 + c \otimes ab \otimes 1) = (Id + S + S^2) \circ \phi^{23}(a(bc) \otimes 1 \otimes 1 + b(ca) \otimes 1 \otimes 1 + c(ab) \otimes 1 \otimes 1) = (Id + S + S^2)(a(bc) \otimes 1 \otimes 1 + b(ca) \otimes 1 \otimes 1 + c(ab) \otimes 1 \otimes 1) = a(bc) \otimes 1 \otimes 1 + b(ca) \otimes 1 \otimes 1 + c(ab) \otimes 1 \otimes 1 + 1 \otimes 1 \otimes a(bc) + 1 \otimes 1 \otimes b(ca) + 1 \otimes 1 \otimes c(ab) + 1 \otimes a(bc) \otimes 1 + 1 \otimes b(ca) \otimes 1 + 1 \otimes c(ab) \otimes 1$.

Similarly,

$(Id + S + S^2) \circ \phi^{12} \circ \phi^{23} \circ \phi^{12} \circ (Id + S + S^2)(a \otimes b \otimes c) = (Id + S + S^2) \circ \phi^{12} \circ \phi^{23} \circ \phi^{12}(a \otimes b \otimes c + b \otimes c \otimes a + c \otimes a \otimes b) = (ab)c \otimes 1 \otimes 1 + (bc)a \otimes 1 \otimes 1 + (ca)b \otimes 1 \otimes 1 + 1 \otimes 1 \otimes (ab)c + 1 \otimes 1 \otimes (bc)a + 1 \otimes 1 \otimes (ca)b + 1 \otimes (ab)c \otimes 1 + 1 \otimes (bc)a \otimes 1 + 1 \otimes (ca)b \otimes 1$.

We now use the axiom of the "weak UJLA structure". □

Theorem 2. *Let (V, Δ) be a weak co-UJLA structure with the co-unity $\varepsilon : V \to k$. Let $\phi = \Delta \otimes \varepsilon : V \otimes V \to V \otimes V$. Then, (V, ϕ) is a "weak (co)UJLA structure".*

Proof. The proof is dual to the above proof. We refer to [6–8] for a similar approach.

A direct proof should use the property of the co-unity: $(\varepsilon \otimes I) \circ \Delta = I = (I \otimes \varepsilon) \circ \Delta$. After computing

$\phi^{12} \circ \phi^{23} \circ \phi^{12}(a \otimes b \otimes c) = \varepsilon(b)\varepsilon(c)(a_1)_1 \otimes (a_1)_2 \otimes a_2$ and

$\phi^{23} \circ \phi^{12} \circ \phi^{23}(a \otimes b \otimes c) = \varepsilon(b)\varepsilon(c)a_1 \otimes (a_2)_1 \otimes (a_2)_2$,

one just checks that the properties of the linear map $Id + S + S^2$ will help to obtain the desired result. □

Theorem 3. *Let (V, η) be a weak UJLA structure with the unity $1 \in V$. Let $\phi : V \otimes V \to V \otimes V$, $a \otimes b \mapsto ab \otimes 1 + 1 \otimes ab - a \otimes b$. Then, (V, ϕ) is a "weak (co)UJLA structure".*

Proof. One can formulate a direct proof, similar to the proof of Theorem 1.

Alternatively, one could use the calculations from [7] and the axiom of the "weak UJLA structure". □

3. Unification of (Co)Derivations and Applications

Definition 4. *Given a vector space V, a linear map $d : V \to V$, and a linear map $\phi : V \otimes V \to V \otimes V$, with the properties*

$$\phi^{12} \circ \phi^{23} \circ \phi^{12} = \phi^{23} \circ \phi^{12} \circ \phi^{23} \tag{10}$$

$$\phi \circ \phi = Id, \tag{11}$$

the triple (V, d, ϕ) is called a "generalized derivation" if the maps d and ϕ satisfy the identity

$\phi \circ (d \otimes I + I \otimes d) = (d \otimes I + I \otimes d) \circ \phi$.

Here, we have used our usual notation: $\phi^{12} = \phi \otimes I$, $\phi^{23} = I \otimes \phi$, $Id : V \otimes V \to V \otimes V$, $a \otimes b \mapsto a \otimes b$ and $I : V \to V$, $a \mapsto a$.

Theorem 4. *If A is an associative algebra and $d : A \to A$ is a derivation, and $\phi : A \otimes A \to A \otimes A$, $a \otimes b \mapsto ab \otimes 1 + 1 \otimes ab - a \otimes b$, then (A, d, ϕ) is a "generalized derivation".*

Proof. According to [7], ϕ verifies conditions (10) and (11). Recall now that $d(ab) = d(a)b + ad(b)$ $\forall a, b \in A$, $d(1_A) = 0$.

$(d \otimes I + I \otimes d) \circ \phi(a \otimes b) = (d \otimes I + I \otimes d)(ab \otimes 1 + 1 \otimes ab - a \otimes b) = d(ab) \otimes 1 - d(a) \otimes b + 1 \otimes d(ab) - a \otimes d(b)$.

$\phi \circ (d \otimes I + I \otimes d)(a \otimes b) = \phi(d(a) \otimes b + a \otimes d(b)) = d(a)b \otimes 1 + 1 \otimes d(a)b - d(a) \otimes b + ad(b) \otimes 1 + 1 \otimes ad(b) - a \otimes d(b)$. □

Theorem 5. *If (C, Δ, ε) is a co-algebra, $d : C \to C$ is a co-derivation, and $\psi = \Delta \otimes \varepsilon + \varepsilon \otimes \Delta - Id$: $C \otimes C \to C \otimes C$, $c \otimes d \mapsto \varepsilon(d)c_1 \otimes c_2 + \varepsilon(c)d_1 \otimes d_2 - c \otimes d$, then (C, d, ψ) is a "generalized derivation". (We use the sigma notation for co-algebras.)*

Proof. The proof is dual to the above proof.

According to [7], ψ verifies conditions (10) and (11). From the definition of the co-derivation, we have $\varepsilon(d(c)) = 0$ and $\Delta(d(c)) = d(c_1) \otimes c_2 + c_1 \otimes d(c_2)$ $\forall c \in C$.

$\psi \circ (d \otimes I + I \otimes d)(c \otimes a) = \varepsilon(a)d(c)_1 \otimes d(c)_2 - d(c) \otimes a + \varepsilon(c)d(a)_1 \otimes d(a)_2 - c \otimes d(a)$,

$(d \otimes I + I \otimes d) \circ \psi(c \otimes a) = \varepsilon(a)d(c_1) \otimes c_2 + \varepsilon(c)d(a_1) \otimes a_2 - d(c) \otimes a + \varepsilon(a)c_1 \otimes d(c_2) + \varepsilon(c)a_1 \otimes d(a_2) - c \otimes d(a)$.

The statement follows on from the main property of the co-derivative. □

Definition 5. *Given an associative algebra A with a derivation* $d : A \to A$, *M an A-bimodule and* $D : M \to M$ *with the properties*

$$D(am) = d(a)m + aD(m) \quad D(ma) = D(m)a + md(a) \ \forall a \in A, \ \forall m \in M,$$

the quadruple (A, d, M, D) *is called a "module derivation".*

Remark 1. *A "module derivation" is a module over an algebra with a derivation. It can be related to the co-variant derivative from differential geometry. Definition 5 also requires us to check that the formulas for D are well-defined.*

Note that there are some similar constructions and results in [9] (see Theorems 1.27 and 1.40).

Theorem 6. *In the above case,* $A \oplus M$ *becomes an algebra, and* $\delta : A \oplus M \to A \oplus M$, $(a \oplus m) \mapsto (d(a) \oplus D(m))$ *is a derivation of this algebra.*

Proof. We just need to check that $\delta((a \oplus m)(b \oplus n)) = \delta((ab \oplus an + mb)) = d(ab) \oplus D(an + mb)$ equals $\delta((a \oplus m)(b \oplus n)) = \delta((a \oplus m))(b \oplus n) + (a \oplus m)\delta(b \oplus n) = (d(a) \oplus D(m))(b \oplus n) + (a \oplus m)(d(b) \oplus D(n)) = (d(a)b \oplus d(a)n + D(m)b) + (ad(b) \oplus aD(n) + md(b))$. □

Remark 2. *A dual statement with a co-derivation and a co-module over that co-algebra can be given.*

Remark 3. *The above theorem leads to the unification of module derivation and co-module derivation.*

4. Modified Yang–Baxter Equation

For $A \in M_n(\mathbb{C})$ and $D \in M_n(\mathbb{C})$, a diagonal matrix, we propose the problem of finding $X \in M_n(\mathbb{C})$, such that

$$AXA + XAX = D\,. \tag{12}$$

This is an intermediate step to other "modified" versions of the Yang–Baxter equation (see, for example, [10]).

Remark 4. *Equation (12) is related to the problem of the three matrices. This problem is about the properties of the eigenvalues of the matrices A, B and C, where* $A + B = C$. *A good reference is the paper [11]. Note that if A is "small" then* $D - AXA$ *could be regarded as a deformation of D.*

Remark 5. *Equation (12) can be interpreted as a "generalized eigenvalue problem" (see, for example, [12]).*

Remark 6. *Equation (12) is a type of Yang–Baxter matrix equation (see, for example, [13,14]) if* $D = O_n$ *and* $X = -Y$.

Remark 7. *For* $A \in M_2(\mathbb{C})$, *a matrix with trace -1 and*

$$D = -\begin{pmatrix} det(A) & 0 \\ 0 & det(A) \end{pmatrix}, \tag{13}$$

Equation (12) has the solution X = I'.

Remark 8. *There are several methods to solve (12). For example, for* $A^3 = I_n$, *one could search for solutions of the following type:* $X = \alpha I_n + \beta A + \gamma A^2$. *Now, (12) implies that* $(2\alpha\beta + \gamma^2 + \alpha)A^2 + (\alpha^2 + 2\beta\gamma + \gamma)A + (2\alpha\gamma + \beta^2 + \beta)I_n - D = 0$.

It can be shown that we can produce a large class of solutions in this way, if D is of a certain type.

Funding: This research received no external funding.

Acknowledgments: I would like to thank Dan Timotin for the discussions and the reference on the problem of the three matrices. I also thank the editors and the referees.

Conflicts of Interest: The author declares no conflict of interest.

References

1. Nichita, F.F. Unification theories: Examples and Applications. *Axioms* **2018**, *7*, 85. [CrossRef]
2. Marcus, S.; Nichita, F.F. On Transcendental Numbers: New Results and a Little History. *Axioms* **2018**, *7*, 15. [CrossRef]
3. Smoktunowicz, A.; Smoktunowicz, A. Set-theoretic solutions of the Yang-Baxter equation and new classes of R-matrices. *Linear Algebra Its Appl.* **2018**, *546*, 86–114. [CrossRef]
4. Motegi, K.; Sakai, K. Quantum integrable combinatorics of Schur polynomials. *arXiv* **2015**, arXiv:1507.06740.
5. Nichita, F.F. On Jordan algebras and unification theories. *Rev. Roum. Math. Pures Appl.* **2016**, *61*, 305–316.
6. Ardizzoni, A.; Kaoutit, L.E.; Saracco, P. Functorial Constructions for Non-associative Algebras with Applications to Quasi-bialgebras. *arXiv* **2015**, arXiv:1507.02402.
7. Nichita, F.F. Self-inverse Yang-Baxter operators from (co)algebra structures. *J. Algebra* **1999**, *218*, 738–759. [CrossRef]
8. Dascalescu, S.; Nichita, F.F. Yang-Baxter Operators Arising from (Co)Algebra Structures. *Commun. Algebra* **1999**, *27*, 5833–5845. [CrossRef]
9. Grinberg, D. Collected Trivialities On Algebra Derivations. Available online: http://www.cip.ifi.lmu.de (accessed on 16 May 2019).
10. Bordemann, M. Generalized Lax pairs, the modified classical Yang-Baxter equation, and affine geometry of Lie groups. *Commun. Math. Phys.* **1990**, *135*, 201–216. [CrossRef]
11. Fulton, W. Eigenvalues, Invariant Factors, Highest Weights, and Schubert Calculus. *Bull. New Ser. AMS* **2000**, *37*, 209–249. [CrossRef]
12. Chiappinelli, R. What Do You Mean by "Nonlinear Eigenvalue Problems"? *Axioms* **2018**, *7*, 39. [CrossRef]
13. Ding, J.; Tian, H.Y. Solving the Yang–Baxter–like matrix equation for a class of elementary matrices. *Comput. Math. Appl.* **2016**, *72*, 1541–1548. [CrossRef]
14. Zhou, D.; Chen, G.; Ding, J. On the Yang-Baxter matrix equation for rank-two matrices. *Open Math.* **2017**, *15*, 340–353. [CrossRef]

MDPI
St. Alban-Anlage 66
4052 Basel
Switzerland
Tel. +41 61 683 77 34
Fax +41 61 302 89 18
www.mdpi.com

Axioms Editorial Office
E-mail: axioms@mdpi.com
www.mdpi.com/journal/axioms

www.ingramcontent.com/pod-product-compliance
Lightning Source LLC
LaVergne TN
LVHW070546100526
838202LV00012B/393